THE HISTORY OF THE FALMOUTH WORKING BOATS

BY ALUN DAVIES

The history of the oyster dredgers on the River Fal, Cornwall. These fishing boats are the world's last oyster dredging fleet to work under sail alone.

10 Claremont Terrace, Falmouth, Cornwall TR11 2AJ

ISBN 0 9515356 0 9

Cornwall Lithographic Printers, Cardrew Industrial Estate, Redruth, Cornwall TR15 1SS

CONTENTS

FOREWARD

Alun Davies was brought up at West Mersea on the East Coast and the sea, boats and oysters have always been very much a part of his life. He left school to join the Seasalter and Ham Oyster Company and it was the oyster beds that they owned on the River Fal that eventually brought him to the Westcountry in 1965, where, to his amazement, he saw for the first time the sailing oyster boats of the Truro River, now, perhaps, more generally known as the Falmouth Working Boats. Although he had grown up among the East Coast smacks and sailing barges he had been completely unaware of this remarkable Cornish fleet, the very last in Europe still working under sail alone. His immediate desire to learn more about the fishery was, however, soon frustrated for he rapidly found that there was very little documented history of the boats or the men that built and sailed them. Resolving to find out more for himself it became a personal quest which has led ultimately to this excellent book.

When Seasalter were bought out by MacFisheries a few years later, Alun turned to dredging himself, first learning the skills by crewing for others, notably Kenny Corke in the *'Evelyn'*, and in 1970 he bought his first oyster boat, the *'Result'*, which was followed in 1978 by one of the real classics, *'Florence'*, built by Ferris in 1895, which he still owns and sails today. His close personal links with the fishery have meant that he has been able to observe and record it intimately over the years, and inevitably much has changed. He was able to talk first hand to many of the older fishermen who had known the dredging in its heyday, and also painstakingly record the dimensions and details of the boats that still survived. Today, 25 years later, many of those old men are long gone and so too are a number of those boats, information that would have been lost forever were it not for Alun's foresight, and combined with the remarkable collection of photographs he has managed to amass during that time he is now able to provide us with this unique social and historical record.

A friend for many years, and a workmate too, it was through Alun that my own interest in the Working Boats was first initiated, first racing with him in the *'Result'* and then for one brief season I had the good fortune to work the oyster beds aboard the *'Florence'*. Although it proved a poor winter financially for pollution and disease had begun to take its toll on this once flourishing fishery, for me there was reward enough in being able to see first hand a way of life that had remained unchanged for over 150 years. In calm or gale, regular as clockwork from 9 till 3, the oystermen set about their work, sails scandalised at startling angles, drifting slowly on the tide, their dredges scraping up the secrets of the hidden banks of the upper Fal, amidst some of the loveliest scenery in the Westcountry, a green and rolling shoreline, an ever changing contrast of water, light and wind. I remember well the great black squalls rising in the north west, bitter onslaughts of sleet and hail that churned the river white and the boats turning gamely to windward in a smother of spray, or those quieter moments, drifting in a glassy calm, barely able to move a dredge, the hours passing in lively banter with the other crews ... and always, lacking an engine there remained the long sail home to Falmouth across Carrick Roads, stiff backed and numb fingered, fumbling to open the thermos for that last lukewarm cuppa. Then, just occasionally, I would gently nag my Skipper, "So when are you going to do something about writing that book of yours?"

"Oh, I'll get round to it one day," he'd mutter, and I'd let it rest at that.

Fortunately for us all, 'one day' he did ...

Mark Fishwick
Falmouth, September 1989.

ACKNOWLEDGEMENTS

The information in this book has been given by so many dredgermen that it would take another book to name them all. However, the following are offered my grateful thanks for the many hours they spent trying to remember old boats and old men to help make this record. Henry Merrifield, Norman Ferris, Jim Morrisson, 'Toby' West, George Vinnicombe, John Merrifield, Bart Moore, Terry Heard, Frank Cock, Paul Williams, Ray Frost and all the other working boat skippers who have helped piece together the jigsaw of the histories of their boats.

Much information was gleaned from the records of HM Customs and Excise and the oyster fishery records of Truro Corporation. Thanks also to Mark Fishwick who wrote Chapter 8 for me when I was running out of steam, and has given me lots of encouragement over the years.

During the twenty or so years spent collecting material for this book, thousands of photos have been collected. Unfortunately it is now impossible to know from whom they all were obtained, but without them the record would be incomplete so thanks to them all. Thanks too for additional photos to Basil Greenhill for the photo of William Ferris (page 4) and his kindness in allowing me to quote from his book 'Merchant Schooners' (pages 4, 7, 31). Also to Truro Museum who provided the photo of 'Mabel' (page 36), Eric Leuchars (pages 127, 128), David Turner (page 164), 'Toby' West for many of the excellent photos of 'Victory' and Bart Moore who allowed me to copy his collection of photos of the many boats he renovated. The front cover photo was provided by Chris Monk.

All the pen and ink sketches were drawn by Keith Harris.

INTRODUCTION

In the far south west corner of England, on the River Fal in Cornwall, can be seen the worlds last fleet of oyster dredgers still working under sail alone. This is not a tourist attraction but a viable commercial enterprise.

During the winter months after the sleek yachts have been laid up, these sturdy vessels will be at work oyster dredging in the great sweep of water in Falmouth harbour known as Carrick Roads, the name deriving from the Cornish word for 'rock'. These unique craft are variously known as 'Restronguet Creek Oyster Dredgers', 'Truro River Oyster Boats' and their more familiar name 'Falmouth Working Boats'. Their story has been more neglected by marine historians than that of any of the many fleets that have fished around the British coast, which is odd as the fact that they alone have survived in their original 19th century rig should merit more than the all too common caption below dramatic photographs in the yachting press "Falmouth Working Boats racing – these beautiful boats which used to be oyster dredgers can be seen racing regularly at Falmouth"! Whilst it is true that the summer racing fleet is now almost entirely composed of boats 'that used to work', there is a larger fleet kept for oyster dredging only which is rarely seen by those to whom the summer racing fleet presents such a familiar sight. During the poorest of oyster seasons this fleet may contract to only half-a-dozen boats, but should the stocks of oysters on the fishery recover to the levels of the late Sixties and early Seventies then doubtless we should see a fleet of over thirty sailboats working again.

What is the secret of the durability of the Falmouth Working Boats? The reason for their continuing existence as a working fleet on the Truro Oyster Fishery is the restrictive nature of the bye-laws forbidding the removal of oysters from the Truro Fishery by means of mechanically propelled craft. By only allowing the fishery to be worked by sailing boats, which have only a fraction of the efficiency of their motorised counterparts, overfishing has been prevented. Ironically it is the dramatic increase in efficiency and catching power that the fishing industry as a whole has enjoyed since it changed from sail to motor power, that now presents the world fisheries with their greatest threat, namely overfishing. That the Truro Oyster Fishery has escaped extinction from overfishing, albeit accidentally by virtue of the bye-laws made in 1868, is our good fortune as it allows us a glimpse of an era long past. An era when Falmouth Working Boats sailed alongside sailing craft of all descriptions from the majestic 'Cape Horners' to the humblest of fishing craft. An era of Tea Clippers, Thames sailing barges, trading schooners, pilot cutters, fishing smacks, collier brigs, lug rigged pilchard drivers, billy boys, nobbys, revenue cutters, Falmouth quay punts and a thousand and one others besides. When the mighty 'J' class yachts raced in the waters of Falmouth, would their millionaire owners have noticed the little gaff cutters that fished for the oysters they doubtless consumed when the days racing was done, and would they have guessed that the same little gaff cutters would still be plying their trade on the oyster banks of the river Fal when their own magnificent vessels were just a fading memory? By the late 1940's virtually all trading and fishing craft had made the transition from sail to motor, and by the end of the 1960's when the last of the thousands of Thames sailing barges, 'Cambria', "hung up her topsail" it was considered by most sail enthusiasts to be the end of the era of sail. Few were aware then, or now, of the existence of a large fleet of sailing boats still at work on the shores of the river Fal.

The boats are typically from 22ft to 30ft long, gaff cutter rig with tall polemasts and long bowsprits. The hulls are decked back to the mast (hence their sometimes being called 'halfdeckers') with narrow sidedecks running aft to a small sterndeck. They are heavy displacement boats with long straight keels which usually have an external ballast keel of about a ton and a half attached, in addition to their two or three tons of internal ballast. In order to work over the shallow oyster banks at low tide, draught is normally limited to about 5ft., beam is about 9ft. They are similar to the Falmouth Quay Punts, with which they are often confused, though the Quay Punts were deeper, carried a mizzen and were used to supply husbandry to shipping, not oyster dredging.

Falmouth Harbour in the days of sail.

Naturally the size of the fleet fluctuates and is at a low level at present time because the fishery has been badly hit by disease for the first time in living memory. Only 6 or 7 boats are working at the moment, but we know from records kept by the licencing authority, Truro Corporation, that it has fallen to this level before on several occasions and can recover very quickly. Just before the First World War catches were steady, but as a consequence of so many fishermen going to the war the beds were given a rest for a few years and during the subsequent 'bonanza' upon their return the fleet increased to 50 sailboats by 1924. (The Falmouth Working Boat is usually called a sailboat on the Fishery in order to distinguish it from the 16ft. oyster punts that also fish for oysters with hand winches, thereby adhering to the 'no engine' rule. These are known as 'punts', 'winkboats' or 'haultow punts'.) At the same time the punt fleet increased to an incredible 187, but of course the fishery could not sustain this size of dredging fleet for very long and declined rapidly over the following years. A similar 'bonanza' occurred after the Second World War, but not quite of the same proportions as that of the 1924/25 season.

The young George Vinnicombe and family out on the 'Boy Willie'. He seems to be more interested in the camera than the beautiful 'J Class' yachts 'Brittannia', 'Westward' and 'White Heather'. 'Boy Willie' has now been converted for dredging.

In those days sail was commonplace in Cornwall (as it was elsewhere in Britain) and fishing craft adapted to whichever type of fishing was most profitable at that particular time, and so during the boom seasons boats of a similar type that were engaged in other work were tempted to fit out for oyster dredging. The fleet might well consist of boats that would more usually be employed as pilchard drivers, longliners, trawlers or maybe even pleasure boats. In short, anything that could be considered remotely suitable would join the fishery and there are several instances of 'dandy rig' (i.e. ketch or yawl), lugsails and even spritsails appearing on the fishery. It is hard to imagine the lugsails and spritsails as being very effective, they certainly found no favour with the oystermen of the Fal who remained faithful to their gaff cutters! The Fishery does not have the benefit of such an influx nowadays, as with the passing of sail for general usage all other methods of fishing are performed by motor boats. These boats are of a completely different shape and would not be capable of sailing even if they were rigged so to do, the result of these limitations being that the fleet will probably never reach the same proportions that it did in the days when every fishing boat worked under sail.

Penzance fishing luggers at Falmouth, it was from fleets such as this many boats were 'imported' to the Fal and converted for oyster dedging.

Many of the oyster boats are now fitted with motors to get to and from the fishing grounds as the Fishery regulations state only that engines may not be used whilst actually fishing i.e. from the time the first dredge is thrown over until the last of the day is hauled back on board. The hours when fishing is permitted are controlled by the licencing authority, 9am-3pm Mon. to Fri., 9am-1pm Sat. At 3pm when the days work is over the engines are all started up and the boats head for home up one of the many creeks on the Fal such as Mylor, Restronguet, Pill, Flushing or Coombe. The picturesque scene formed by the oyster fleet under sail is transformed in moments as all sail is lowered and everyone motors home, save only the few remaining boats without an engine who must sail home or scrounge a tow should the wind have fallen light. Several of the boats have opted for an outboard motor as they can be taken inboard when not in use and so do not have the penalty of dragging a propeller as is the case for the bulk of the fleet who prefer to have inboard diesels of 15-30 horse power.

The control of the oyster beds has now passed to Carrick County Council who swallowed up Truro City Corporation during a local government reshuffle, but this has had little effect upon the running of the Fishery. As well as setting working hours they control the size of oysters to be caught. This is presently 2 ⅝″ diameter and is measured by means of the 'oyster ring', a gauge provided to the dredgermen by Carrick County Council. Any oyster that can be made to hang on the ring without falling through is considered to be legal. A bailiff is employed by Carrick in order to ensure all the regulations are fully observed, and each day he goes on board the boats to check the catch.

That little has been written about this fleet is partially due to the lack of records. The Truro Corporation did keep records of the licences issued, details of the boats, numbers of oysters taken from the Fishery annually as well as other information relevant to the Fishery, however, all the records prior to 1914 were lost in a fire in the Truro Town Hall that year. Some of the history of the boats in this book was obtained through the Customs and Excise records, but the bulk of the information here has had to be gleaned from the long memories of the older dredgermen, and so whilst every attempt at accuracy has been made, some of it can only be as precise as the memory of the person from whom it was obtained. In the case of conflicting opinions a decision has to be made based on all the facts that are gathered, and hope it is as near the truth as possible. It should be remembered that whilst the information given about a particular boat may be given in the best of faith, accuracy cannot always be guaranteed as in the absence of records it may have been obtained from someone who did not have the correct information.

It cannot be emphasised too often that it is due directly to the bye-laws prohibiting the use of engines that we are still able to boast a fleet of fishing boats in Great Britain that still work and earn their living under sail alone. Were the bye-laws repealed and the beds opened to power boats, they would be cleaned out in no time as we have witnessed in our other oyster fisheries in this country. If there is no wind the boats cannot sail and the fishery gets a rest, if there is too much wind again the beds may get a rest, so that even the most efficient of dredgermen in a sailing boat will only catch maybe half the quantity he would catch were he in a motor boat, probably far less. The bye-laws may not have been created with the long term future in mind, but no one can fail to see their effectiveness on the Truro Oyster Fishery.

Truro Corporation oyster ring.

CHAPTER 1

The Boatbuilders of the Fal.

The boats of the Falmouth Working Boat fleet differ to those of other fleets as the majority of them were not built purposely for oyster dredging, but were 'borrowed' from a variety of fleets. In the case of most fleets of sailing boats, the vessels would be pretty much identical as they would have evolved into the optimum shape for the type of work and area, making each boat's origin instantly identifiable. However in the case of the Falmouth Working Boats, only a handful were built exclusively for oyster dredging, the remainder coming not only from other Cornish fleets but as far afield as Norfolk, Lancashire, Hampshire, Devon and even Germany. Just so long as they had the right characteristics of length, draught and beam any boat would be considered for use as a Falmouth Working Boat. To the uninitiated it may sound as if they are all identical if these three vital statistics are the same but this is far from the truth as can be seen from many of the photos in this book taken whilst the boats are ashore. There are great variations in hull shape within the fleet which is not normally to be found in the fishing boat fleets of Great Britain, which whilst adding interest to their history does pose the problem of defining precisely what a Falmouth Working Boat should be.

These stern views of 'Happy' (left), 'Result' (middle) and 'Leila' (right) illustrate the great variety of hull shapes.

Because of this diversity of origin it is impossible to define a Falmouth Working Boat by shape, and perhaps the only definition can be that the boat has actually dredged oysters on the Truro Fishery. It can become confusing for though they have been called 'Restronguet Creek Oyster Dredgers', and 'Truro River Oyster Boats', they might

Happy

Result

Leila

not have been built at Restronguet or Truro and even though they are now called 'Falmouth Working Boats' they have less connection with Falmouth than they have with either Truro or Restronguet. However, as 'Falmouth Working Boat' is the most popular name (few local people have heard of the other two names) and as it is the title used by the committee organising the summer racing, it is probably the one that will survive. The builders of many boats will now never be known as in many instances we don't even know their port of origin, but the following are the most famous of the local shipwrights whose boats were built as oyster dredgers.

The best known of the local boatbuilders were William 'Foreman' Ferris, Peter Ferris, Frank Hitchens, Tom Hitchens, Stephen Brabyn and more recently Terry Heard. The most famous of these was undoubtedly William 'Foreman' Ferris of Restronguet. His first boat was the 'Harriet' built in 1861 at Low Hill in Pill Creek. The second was the 'Six Brothers' in 1890 at Ponsmain, the same place he built the 'Florence' 5 years later in 1895. He built many other craft besides these Working Boats and perhaps his finest was the beautiful 101 foot coasting schooner 'Rhoda Mary' in 1819, claimed to be the fastest of her kind ever built. Though appearing to be a dunce at school he was considered to be a near genius by many at the time of his death, and his feats of shipwrighting are still legendry. Tales are told of how he put a new carvel bottom in the clinker built 'Ziguener' to the astonishment of everyone as it was not considered possible. It was done in order to reduce her draught and make her suitable for oyster dredging, and he did it by rigging sheer legs on the beach and cutting her right through on the waterline. His great rival Frank Hitchens said it was impossible, but he did it and for a cost of only £22 including the iron keel. He is said to have planked one side of the 'Harriet' in a day, though it may have been the 'Florence' as in the opinion of local shipwright Tony Murray who has just finished major renovations on 'Florence' her whole design seems to have been planned with speed of building very much in the forefront. The Ferris family always relate with glee how Frank Hitchens had already split three planks attempting to coax a particularly difficult plank round the transom of one boat, when 'Foreman' just happened to come along, and showed him how to do it at the first attempt! Between these two master shipwrights there was great competition and although 'Foreman' Ferris was thought of as the better craftsman, Frank Hitchens had a reputation for building faster boats.

It was said that Foreman's wife could build as well as he could, and that the two of them went to a field with a cartload of timber, several pasties and a bottle of rum. She built one side of the boat, he the other and the two were joined by bolting them together down the keel!! It certainly makes a good story even if it is a little improbable.

In fact another Ferris, also called William but who lived in Flushing, built a punt in two halves which he then joined by bolting them together. It may be here the stories got a bit mixed up!

William 'Foreman' Ferris. To quote Basil Greenhill in his book 'Merchant Schooners', "William Ferris was a quiet and steady man possessed of a genius for the use of tools and for the rightness of the feel and shape of those things which he made with them."

'Foreman' Ferris was a man of great patience and loved any sort of work with wood. He carved a riding stick which was valued at £130 by the Royal Cornwall Society, a great deal of money at that time. His mantlepiece was carved by him and included knots of all kinds and acorns. Even the tillerhead on the 'Rhoda Mary' had decorative roses which he carved in a few hours. He and his younger brother Peter lived at Pill Creek where his father was also a master shipwright and foreman of the shipyard owned by Mr. Hodges and later Hitchens and Ford. It was at this yard at Pill Creek where the brothers both started their careers and that 'Foreman' Ferris built 'Harriet', though this may have been as a private venture as the main shipbuilding at the yard was of coastal schooners.

In the late 1850's J.Stephens started a shipyard at Yard Point beside the sawpits, and in 1867 'Foreman' Ferris became his head man and moulder. The yard then started to improve and output stepped up with one schooner launched each of the next ten years – all designed by 'Foreman' Ferris – the first being 'Rhoda Mary' in 1868. It was 'Foreman's' proud boast that he could draft, draw and scale, and was thus involved in his boats from conception to completion. Not a man to rest on his laurals, he was constantly striving to improve his hull forms, and one of his favourite ways of doing so was to build model boats which he sailed in local ponds. It was from his observations of these that many improvements to his designs were to follow.

It is unclear as to exactly what happened with regard to the yard. 'Foreman' Ferris married J. Stephens daughter and was apparently left the yard by Stephens. It closed down probably in the 1880's which may be the reason for both the 'Six Brothers' and the 'Florence' being built on the opposite side of the creek at Ponsmain during the 1890's. This was where he also built his last boat, a punt called 'Granny and Grandfather', at the age of 87. He died peacefully in his chair as he sat talking and rolling tobacco for his pipe.

Amongst the other boats 'Foreman' built were three 18′ gaff cutters 'Twinkle', 'Buttercup' and 'Zephyr'. They were not built as dredging boats but were of the same shape, the 'Zephyr' being a scaled down version of the 'Florence'. The 'Twinkle' proved to be so fast that she regularly competed in the 28′ class when racing in the regattas! Smaller than the 18′ boats were the 'Tar Baby', 'Whim' and the 'Tom Tit' which is still racing in Falmouth. He also built many yachts designed by Harley Mead, of which one may have been the 'Little Windflower', and another was definitely the cutter 'Merlin'. She was 30′ long, 10′ beam with a draught of 6′6″ and was said to be very similar to 'Florence'. She was built either for a lawyer called Mr. Bennetts or T.C.M. McGuinn of Guernsey.

The similarity in hull shape to 'Florence' can be seen in this photo of 'Zephyr', also built by William Ferris.

'Twinkle' with spinnaker set, though little wind to fill it.

Peter Ferris like his brother William left the yard at Pill Creek to go and work at Yard Point. To quote from Basil Greenhill's book 'The Merchant Schooners', Peter Ferris was "a man of some independence of disposition, of marked sense of humour and inclined to be erratic in his working methods. Like his brother he was illiterate, and like him also had begun to work as soon as he was able in the yard at Pill Creek". He was probably overshadowed by his elder brother who was of course his foreman, and perhaps he decided for this reason to leave Yard Point to seek his fortune elsewhere. He left in 1870 and by the following year was headman at the boatyard in Charlestown Dock, Charlestown. This was owned by Thomas Stephens, who was no relation to the other Stephens for whom his brother was foreman, but they did in fact become partners as Thomas Stephens was a business man, not a shipwright, and so John Stephens managed the yard whilst Thomas Stephens attended to the paper work. They built a barquentine here at Charlestown, but it was when they moved to another yard in Fowey harbour (on the site the Royal Fowey Yacht Club now occupies) that he started to build schooners that matched his brothers. The most famous of these were the 'Katy Cluett' and the 'Ocean Swell'. By now the brothers had achieved the reputation for being the builders of the fastest schooners on the coast and we will never really have the answer to the question which was the faster of the two boats, 'Rhoda Mary' or 'Katy Cluett'. They only met once in their careers, it was off Falmouth and the 'Rhoda Mary' showed 'Katy Cluett' a clean pair of heels, but it was a pity that they did not meet a few more times in order to be certain that this was not a fluke.

However, in 1878 J. Stephens closed his yard at Fowey, and then Peter Ferris moved to Looe where he started building fishing craft. It was here that he constructed a wide variety of craft including Falmouth Working Boats, fishing luggers and yachts. He built four Working Boats for the Truro Oyster Fishery, the first 'Serica' (1912), then 'Katherina' (1913), the 'Five Sisters' (1914) and finally 'Tern' (1920). A fifth boat built by him in 1904 was 'Nellie', which though it raced with the Falmouth Working Boats at Falmouth never actually dredged. His boats were all very distinctive having a rounded cutaway forefoot, though not 'Nellie' which was built as a crabber. He also built two yachts, the first being the 27′ 'Pearl' in 1902 (later to become 'Preal'), and a gaff cutter called 'Mariner A' in 1912. His most famous yacht, however, was not built as a yacht, but as a fishing lugger in 1896 called 'Lily'. She was to find fame as the 'Moonraker of Fowey' after she was purchased in the 1930's by Peter and Anne Pye, who wrote of their extensive cruising in her in their books 'Red Mains'l', 'The Sea Is For Sailing' and 'A Sail in a Forest'.

'Nellie' racing at Falmouth whilst owned by Ben Collins.

Frank Hitchens was a pupil of 'Foreman' Ferris at Yard Point and learnt his trade from him. He built his boats on Yard Point after Stephens yard had closed, but he built his first there in 1868 (this was 'Shadow') when Ferris was starting as foreman. It was almost certainly due to 'Foreman' Ferris that got Frank Hitchens interest in Working Boats became aroused, and after the 'Shadow' he went on to complete four more. These were the 'Royal Oak' (1884), 'Mabel' (1895), 'Evelyn' (1898), and his last 'Softwings' (1900). He built on the beach with his son Tom, and it may even have been Tom Hitchens that built the 'Softwings'. Frank's best boat might well have been the 'Evelyn' which he built for his son, with the specific aim of building a boat to beat all others. He may well have succeeded in this, but we will never be able to judge for ourselves as fast though 'Evelyn' undoubtedly is, we will have no opportunity of comparing her with 'Mabel' who left the port in the early 1930's leaving behind a reputation as that would certainly have made a race between 'Evelyn' and 'Mabel' anything but a foregone conclusion!

Stephen Brabyn's boatyard was halfway up Calenick Creek on the port hand side. Using wood supplied by Truro Timberyard, sawn at Scorrier Mill, he built 'Queen of the Bay' (1892), 'Echo' (1897), Winnie

'Temptress' built by Brabyn and veteran of Atlantic crossings.

(1898), Mildred (1898), Beatrice (1903), and 'Ivy' (1905). In addition to these Working Boats he also built several barges and the yacht 'Temptress' later to be made famous by Allcard for his book of his single handed transatlantic crossing with her. Stephen was probably a relation to the Brabyn's at Padstow and Looe who were also boatbuilders.

It is unusual, to say the least, when writing up the history of a fishing boat fleet to be able to include recent boatbuilding, but in the case of the Falmouth Working Boats we can. As the cost of building new boats out of the traditional material, wood, has become prohibitive nowadays, the dredging fleet faced difficulties in replacing its ageing boats, many of which were approaching 100 years of age. That is until Terry Heard decided to try and build Working Boats in fibre-glass at his Tregatreath Boatyard.

Terry, who sadly died in 1985 aged only 57, was an immensely popular man, and any visit to his yard at Tregatreath where he worked with his brother Colin and son Martin, would certainly brighten up even the greyest of days. Whilst obviously in business for profit, it was quite clear that a little detail like that would not be allowed to interfere with the far more serious business of enjoying life to the full.

Terry Heard and son Martin stand at the bow of a new fibreglass dredging punt, whilst brother Colin rests on his adze at the stern. Behind them stands a new GRP 28 footer which was doubtless the occasion for which this photo was taken in 1978.

Terry built wooden Working Boats before turning to fibre-glass, the first being 'Result' in 1961, then 'St. Melorus' in 1968. Two others were also built as dredging boats, 'Chrystal Spring' and 'Murre', David Cain working 'Chrystal Spring' and Peter Divers, 'Murre'.

Terry Heard cramping up a plank on 'St Melorus'.

After completing the four wooden boats, Terry saw the potential in fibre-glass, and with his wooden 'St. Melorus', designed by Percy Dalton, he made the mould for the first of the 28ft GRP boats. This was in 1970 and he called the boat 'Melorus' and kept it for his own use. The next was not built until 1973 (probably the longest period the yard has been without one under construction) which was sold to dredgerman John Walls of Flushing, who with the help of shipwright and dredgerman John Moore fitted her out for oystering. He called her 'Three Sisters' as he has 3 daughters. It proved to be a great success and soon many others followed suit, 'Kindly Light' for John Moon, 'Helen Mary' for Ron 'Hammer' Laity and 'Kayte' for Clive Warrilow. Others were built as cruising yachts and in 1978 local doctor Phil Slater had 'Rebecca' built purely for racing. In 1979 the mould was altered (rendering the boats from the first mould a limited edition!) to a new improved design of Percy Dalton's. The result of this has only provided one dredging boat so far (due to the decline in the oyster fishery over the last few years), and this was the 'Orca' which John Moore fitted out for himself after selling the 'Royal Oak'. It continued to dredge until it's next owner, Mick Rowan, sold her to a Frenchman who put a cabin on her and took her back to France as a cruising yacht.

To fill the demand for a smaller single handed dredging boat,

Percy Dalton designed another boat for Terry, this time 23ft. These too have proved themselves to be excellent dredging boats as well as very competitive for racing. The first of these was 'Varuna' (owner J. Allen) in 1978 and was soon followed by 'Carrie' (K. Corke), 'Girl Sara' (J. MacDonald), 'Lena' (T. Arthur) and the last one so far to have been built for dredging was fitted out by B. Lewarne of Coombe Creek.

Though at present this represents the full list of boats built for dredging from Terry's yard, his son, Martin, has taken over and is busy fitting out 28ft and 23ft boats as yachts. They have proved to be very popular as cruising boats as they are both seaworthy and spacious, their seaworthiness being proved by John Jackson who bought the 'St Melorus' from Terry, fitted a cabin, decked her in, and then sailed her to the West Indies.

There were no other dredging boats built by boatyards on the Fal, but several have been built by local men as a 'one off' or perhaps in the case of Norman Ferris we should say 'two off' as he built both 'Zephyr' and

Mike Parsons' 18 foot 'Thelma' at Mylor Regatta.

'Gazelle'. Norman was the grandson of 'Foreman' Ferris and built many fine punts on Devoran Quay where he lived. He built the 'Gazelle' first about 1940, and the 'Zephyr' in 1958.

In 1963 Mike Parsons built the 'Leila' to replace his previous craft 'Thelma' which was lost in a gale. A sad loss to the Working Boat Fleet was the lovely 'Six Brothers' which broke adrift in a S.E. gale and was wrecked on Greatwood Quay. To replace her George Vinnicombe went to Skentleberrys at Plymouth who constructed a new 'Six Brothers'

'Rhoda Mary' (foreground) alongside 'Shadow'. Roy and Rex Collins took her lines as a model for their new boat.

The much loved 'Six Brothers' (the 'Old Six') lies a total wreck off Greatwood Quay, having broken her moorings in a SE gale.

from lines drawn up by Percy Dalton. Peter Mays and Alf Smithers built the 'Alf Smithers' at Penryn, Peter having to complete her himself at Cowlands Creek after Alf's death. 1979 saw the launch of the beautifully built 'Rhoda Mary', a joint venture by brothers Roy and Rex Collins who had previously been dredging in Frank Vinnicombe's 'Boy Phil'. 'Rhoda Mary' was built at Mylor of local larch from Enys Wood and named after their mother, not the William Ferris schooner as is usually assumed. The last wooden boat to be built was 'Mabel', though not strictly speaking a Working Boat as she has not been built to work, only race. She still deserves a mention as she has been built on the lines of the original 'Mabel' built by Frank Hitchens, obtained by Ralph Bird, well known locally for his work in restoring gigs. The only one to have been built of ferro-cement is 'Agnes', another Percy Dalton design. Percy Dalton had a major influence on many of the boats built during the last 20 years and was still trying to improve his designs up until the time of his tragic death in a house fire in 198 .

Although he never actually built a Falmouth Working Boat himself, Bart Moore has surely earned his inclusion in this chapter as he saved so many of the boats from an untimely death. He has performed 'major surgery' on several of them on the beach below his house on the bank of Mylor Creek, and repair to a lesser extent on many others. One of the largest operations he undertook was on the 'Florence' which he practically rebuilt in 1955. Other craft which he has worked on include 'Rippling Wave', 'Six Brothers', 'Lady Beatrice', 'Boy Peter', 'Gypsy', 'Swallow', 'Lily', 'Winifred', 'Royal Oak' and 'Leila'.

Bart Moore at work on the 'Boy Peter'. Not the most common shape for a dredging boat!.

Other boats to have been saved by Bart Moore include 'Petrel' (below) and 'Winifred' (above).

CHAPTER 2

A Typical Working Boat

Let us take a look at a typical Working Boat, a good example of the type being 'Florence' built by William 'Foreman' Ferris in 1895. She is 28′9″ in length, has a beam of 8′9″ and a draught of 5′10″. She is full in the bilge with low floors and has her maximum beam on the waterline. The full bilge allows her to lay on her side without legs and to rise again on the tide should she need to do so, as is the case with the boats kept in Pill Creek and the upper reaches of Mylor and Restronguet Creeks where the boats dry out completely at low tide. Her forefoot is only slightly rounded with a draught forard of about 4ft., and she has a long straight external ballast keel of approximately 25cwt. It is about 20ft long 8″ deep and 6″ wide and in addition to providing the necessary stability is of great benefit in maintaining a true course when the helm is left unattended, unlike short keeled boats which have a tendency to yaw and broach and need more attention on the helm. The entry is quite fine and the run is long and clean with a high tucked transom stern, the bottom of which meets the waterline. The stem is almost upright but the top inclines slightly forward, the stern post has a greater incline though this time the top leans aft. The rudder is hung outside on the sternpost in the usual manner for craft of this type and is supported by means of a gudgeon and pintle top and bottom, with the bottom of the rudder having a pin which fits down into a hole in the keel skeg.

'Florence' showing her lines at Fox's Boatyard.

As she was built as a multi-purpose vessel in order to be able to earn a living in the summer months when the oysters were out of season, she had to be capable of working out at sea in bad weather, not just in the sheltered waters of the river Fal. She has a good sheerline and good freeboard and was able to go to Plymouth to work on the whiting fishery there and was known to have even ventured as far afield as the French coast, though times must have been pretty hard to warrant risking a trip like that in an open boat even in the summer. Generally a boat such as this would only be expected to do a little trawling, handlining for mackerel or longlining in the comparatively sheltered Falmouth Bay.

View of the new steel floors and laminated frames during the rebuilding of 'Florence' in 1987.

Her hull is carvel built and the planking is local larch 1⅛″ x 6″ with frames originally being local oak cut from natural turns 2″ x 2″, were 12″ apart and in one piece down to the keel. The floors too were of local oak, about 3″ wide and fastened down through the keel in the manner usual for wooden boats of this type. These have all been replaced now with a rather unconventional technique proposed by shipwright Tony Murray. When in 1986 a survey found that nearly every frame and floor had been rotted out by fresh water trapped in behind them, his suggestion was that they be replaced with frames laminated in iroco and floors of steel. This method offered the dual advantages of economy and greater strength, the economy from the labour savings using modern techniques like lamination and the strength from the use of steel for the floors (which now rise to the waterline instead of the garboard and two planks above it only) as well as the superior strength of the laminated frames over the original wooden ones. At the chainplates where the stress induced by the mast and rigging is at it's worst, the floors are connected by steel straps to the chainplates so the weight on the chainplates is taken up by strap that is bolted through the laminated frame, both of which are bolted to the steel floor which is itself bolted through the frame and down through the keel. The foot of the mast is fitted into a steel plate welded across the floors and so once again the considerable strain that is normally taken by a small area and frequently leads to problems with leaking garboards is spread over a large area to reduce stress.

For longitudinal strength the boats have a beam shelf running the full length of the boat, and fastenings are normally iron, though in the oldest of them the old method of using trenails is still to be found. These are wooden pegs that are hammered tightly into a hole that is drilled in the plank and frame, once in the water swells the trenails and the resulting fit is as tight and strong as could be achieved by any other method of fastening. The only danger is that should the boat be allowed to dry out on the beach the trenails may contract again and fall out! This type of fastening, which is also known as 'trunnelling', could be seen on the 'Morning Star' and 'Susan Jane'.

There is always an exception to every rule, and here 'Maiden Nellie' has a bowsprit on the starboard of her stem!

There is a focs'l under the foredeck and though there is no reason for it to be so placed, the entrance seems to be on the port side on all of the boats in the same way that the bowsprit is traditionally on the port side of the stemhead. The fo'c'sle is normally used only as a storage locker for sails etc. as there is little need for accommodation, though 'Florence's' must have seen some use as accommodation when she went to work on the whiting off Plymouth. Profits would have been unlikely to be sufficiently large to stretch to accommodation ashore.

Floorboards are loose and lay on bearers which are positioned to put the sidedecks at waist height as this is the best height for working the oyster dredges. The internal ballast is laid out under the floorboards and in 'Florence' is approximately 4½ tons of pig iron. Unusually she has a lower 'floor' of iron ballast specially cast to the shape of her bilge, they fit around the floors and frames leaving a flat base upon which the rest of the ballast sits. Each of these must weigh about 2cwt. and there are about 16 of them in all, 8 on each side of the keel. Presumably the idea behind going to this amount of trouble was to improve the stability and speed by getting the ballast as low in the boat as possible. Nowadays many of the owners try to replace the iron with lead for the same reason, though it is a lot more expensive.

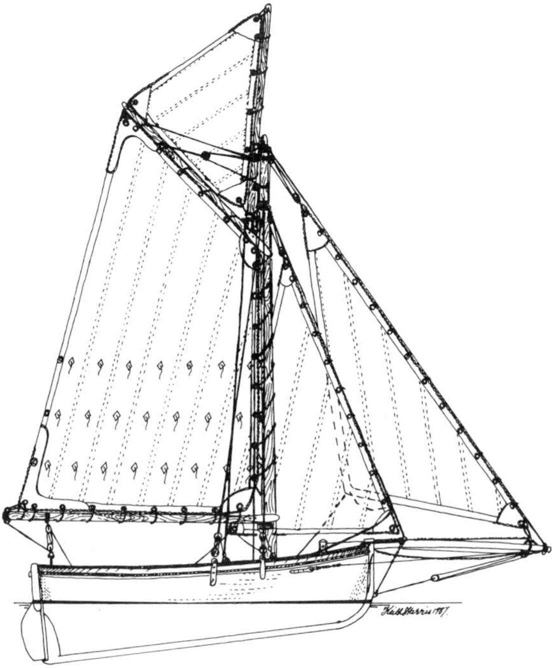

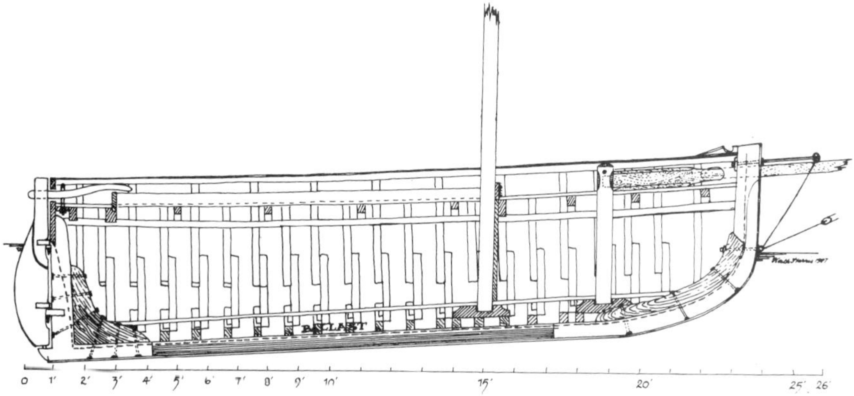

The mast is 39ft overall, 8″ at the base tapering to 5″ at the masthead. It is local pine from Enys Wood and sits on the keel with a support band at deck level. The rigging is simple, comprising only of two main shrouds either side and a forestay. Backstays are carried for racing but are not necessary on a stout spar like this for dredging. The forestay on a dredging boat (as opposed to a boat intended only for racing) comes down on to the stemhead, passing back through a hole in the stemhead to the foredeck where it is made up. The purpose of this is to allow the staysail to be lowered flush to the deck when dredging otherwise were there a bottlescrew or lanyard on the stemhead the sail would not be able to get down the last 9″ or so, and would have the unwanted effect of keeping the sail full of wind when fully lowered. On a particularly tall mast, or a mast a bit on the light side, crosstrees might well be fitted.

'Florence' in her dredging rig.

For many years 'Florence's' mast was 44ft 6″ and needed crosstrees, but now it has been shortened to 39ft they have been done away with. The other spars are the gaff (19ft), bowsprit (26ft) and boom (22ft 6″) and again are like the mast solid spars of local pine. The mast is upright, the bowsprit projects outboard about 15ft parallel to the water and the boom extends 6ft over the transom giving the boat a graceful, old fashioned appearance. 'Florence' carries three sails whilst working, mainsail, staysail and jib, all of which are terylene which can stand far more abuse than the canvas sails which went out of common use on the Fishery some twenty years ago. The mainsail is made of 10oz terylene, the luff 17ft, leech 36ft, foot 21ft 6″, head 18ft 6″. The staysail is 20ft luff, 18ft leech, 7ft foot and the jib is a high cut jib with a 25ft luff. In heavy weather a storm jib with a luff of only 6ft is set and is set flying as are all the jibs even the big racing ones. Her winter rig is only about 250 sq.ft. but this is increased to about 1,000 sq.ft. for racing. Standing rigging on 'Florence' is heavier than most, being plough steel, unlike that of many boats who favour the lighter stainless steel. Another less usual feature on 'Florence' is that lanyards are used on the shrouds in place of the more common bottlescrews, as they can be set up equally as hard

'Florence' in her racing rig behind 'Zephyr'. 'Leila' in background.

as bottlescrews with the added advantages that they are maintenance free, cheap, never seize up and should one need replacing quickly there is always a lashing around to get you out of trouble. Her deadweight is 10 tons with a registered weight of 6 tons.

Bart Moore rebuilding 'Florence' in 1955.

As well as the sails, the running rigging is also all man made fibre, but the blocks are still of wood as they have a longer life span than the modern blocks and are generally more suitable. Upon reading this description of a Falmouth Working Boat, no doubt some of the gaff rig enthusiasts will feel that they are spoilt by the use of all these modern ropes and sails. What must be remembered is that these craft are not a fleet of 'Old Gaffers' in the same sense as the Colchester oyster smacks or Whitstable yawls, for as those fleets finished working many years ago their renovators must use the materials that were in use at the time they ceased working if they are to be true examples of their type. At that time the materials in common usage were canvas, hemp, manilla etc. and so should be used for an authentic restoration of boats of that era. However, this is still the era of the Falmouth Working Boats and so it is quite 'legitimate' for them to use terylene sails, man-made ropes, hollow spars, alloy spars, sheet winches whilst racing and anything else that can reduce the cost or labour needed to run a fishing boat in the 1980's, as doubtless would the Colchester and Whitstable boats were they still working. Though were the Fal Oyster Fishery to close down tomorrow then of course we could not develop them any further than their present state if they were to remain true examples of their type.

CHAPTER 3

Boat Histories

Since about 1870 some 210 boats have been involved with the fishery, and probably quite a few more besides. All the boats listed in the Appendix have been collected from fishery records, customs records and local dredgermen with long memories, but undoubtedly many have been left out.

Of the 210 boats discovered to have dredged on the fishery, the port of building of 111 of them only is known. The port of Porthleven claims the highest percentage of these, 22 having been built there, next Mylor with 17 and then Restronguet with 16.

Mevagissey	14 boats
Looe	9 boats
Falmouth	10 boats
Calenick	5 boats
St. Ives	5 boats
Mousehole	5 boats
Gorran Haven	4 boats
Penryn	2 boats
Newquay	2 boats
St. Mawes	1 boat
Cowlands	1 boat
Gerrans	1 boat
St. Just	1 boat
Helford	1 boat
Porthallow	1 boat
Itchin Ferry	1 boat
Plymouth	1 boat
Polperro	1 boat
Morecambe	1 boat
Kings Lyn	1 boat
Germany	1 boat

Of these forty-eight were built specifically for dredging and the rest adapted for the work. The following table shows the boats from each area, the boats built for oyster dredging *are in italics* and those still in commission as yachts, fishing boats or dredging boats are marked with an asterisk.

NAME	PORT OF BUILDING	DATE	BUILDER	DIMENSIONS
Ailsie	Porthleven	1882	Unknown	43′ x 13′ 6″ x 6′ 75″
Anna	Unknown	(1869)	Unknown	40′
Alfred	Unknown	(1869)	Unknown	24′
Alarm	Unknown	(1869)	Unknown	27′
Ann (ex Georgina)	Unknown	(1902)	Unknown	26′ 7″ x 8′ 9″ x 4′
Agusta	Unknown	(1869)	Unknown	25′
Augusta	Unknown	(1871)	Unknown	18′ 6″ x 6′ 6″ x 3′
Agenora	Unknown	(1869)	Unknown	22′
*Alf Smithers**	Penryn & Cowlands	1972	A. Smithers P. Mays	26′
Boy Ben	Unknown	(1869)	Unknown	24′
Boy Peter	Mousehole	1905?	Unknown	26′
Brothers	Unknown	(1891)	Unknown	34′
Blanche	Falmouth	(1922)	Unknown	22′ 5″ x 8′ 4″ x 2′ 8″
Boy Jim (ex Ellen)	Gorran Haven	(1894)	R.D. Pill	30′ 7″ x 10′ 7″ x 4′ 4″
Blue Eyed Maid	Unknown	Unknown	Unknown	Unknown
Boy Warry	Porthleven	1895	Unknown	28′ x 8′ 4″ x 5′
*Blossom** (ex Kindly Light)	Mylor	1973	T. Heard	28′
Boy Willie*	Porthleven	Unknown	Unknown	29″ 6″ x 9′ 3″ x 4′ 1″
Cock O'The Walk	Unknown	Unknown	Unknown	Unknown
*Chrystal Spring**	Mylor	1966	T. Heard	20′ x 8′ 6″ x 3′ 9″
Charm	Porthleven	1899	Unknown	29′ 9″ x 9′ x 5′
Crest of the Wave	Unknown	Unknown	Unknown	30′
Crested Wave	Unknown	(1879)	Unknown	29′
Caroline	Unknown	(1869)	Unknown	45′
Cheyney Rock	Unknown	(1877)	Unknown	28′
Chance	Restronguet	1903	Wm. Ferris	22′ 6″ x 8′ 4″
*Carrie**	Mylor	1979	T. Heard	24′
Catherine	Unknown	(1875)	Unknown	27′
Deerfoot (ex Xerifa)	Feock	(1869)	Unknown	28′
Dawn	Unknown	(1880)	Unknown	20′
Dorothy Louise* (ex Momento)	Porthleven	1897	Oliver	31′ 9″ x 10′ 9″ x 5′ 6″
Dolphin	Unknown	(1869)	Unknown	36′
Dolphin	Unknown	(1877)	Unknown	22′
Dairy Maid	Unknown	(1924)	Unknown	34′ x 10′ 6″ x 5′ 4″
Deacher	Cowlands	Unknown	Unknown	27′ 5″ x 8′ 3″ x 4′ 9″
Echo	Unknown	1896	Unknown	23′
Endeavour*	St Ives	1911	Unknown	26′ 6″ x 9′ 4″ x 4′ 4″
Emma	Restronguet	(1869)	Ferris Hitchens	30′
Ellen	Unknown	(1870)	Unknown	26′

NAME	PORT OF BUILDING	DATE	BUILDER	DIMENSIONS
*Evelyn**	Restronguet	1898	F. Hitchens	27′ 3″ x 9′ x 4′ 6″
Elizabeth Mary*	Looe	1909	R. Pearce	
Ellen	Mevagissey	(1870)	Unknown	28′ 3″ x 8′ 5″ x 5′
Eaglet	Unknown	(1895)	Unknown	32′
Edith Annie	Falmouth	1884	Unknown	30′ 5″ x 9′ x 4′ 4″
Elizabeth	Unknown	(1869)	Unknown	17′
Express	Unknown	Unknown	Unknown	Unknown
Eliza	Unknown	(1869)	Unknown	26′
Etta	Unknown	(1869)	Unknown	21′
Ergo	Unknown	(1869)	Unknown	26′ 3″ x 8′ 8″ x 4′ 7″
Edith ("Galloping Donkey")	Mevagissey	Unknown	Unknown	22′ – 24′
*Emma Louise**	Mylor	1979	T. Heard	24′
*Florence**	Restronguet	1895	Wm. Ferris	28′ 9″ x 8′ 9″ x 5′ 10″
Five Sisters (Dolly)*	Looe	1914	P. Ferris	24′ 5″ x 8′ 3″ x 3′ 7″
Fearnot	Porthleven	(1881)	Unknown	28′ 5″ x 9′ 2″ x 4′ 5″
Flora	Unknown	(1869)	Unknown	26′
Fal	Unknown	(1869)	Unknown	24′
Fairy	Gorran Haven	(1869)	R.D. Pill	25′ 6″ x 7′ 7″ x 3′ 9″
Forget Me Not	Unknown	Unknown	Unknown	Unknown
Fairy Queen	Unknown	(1871)	Unknown	21′
Five Brothers (ex Alien)	Restronguet	1901	Wm. Ferris (son)	26′ 7″ x 9′ 7″ x 4′ 6″
Flying Cloud*	Restronguet?	Unknown	Wm Ferris	16′ 6″ x 6′ 9″ x 2′ 9″
*Gazelle**	Restronguet	1940	N. Ferris	
Glory	Unknown	(1869)	Unknown	30′
George Glasson*	Porthleven	1898	Oliver	30′ 8″ x 10′ 2″ x 5′ 9″
Gypsy* (ex Chaidee)	Porthleven	1900	Unknown	25′ 3′ x 7′ 10″ x 5′ 24′
George & Mary	Unknown	(1869)	Unknown	24′
Girl Sarah*	Mylor	1983	T. Heard	24′
Honour*	St Ives	Unknown	Painter	23′ 8″ x 7′ 8″ x 3′ 9″
*Happy**	Helford	1896	Unknown	25′ 2″ x 7′ 7″ x 4′ 2″
Heartsease	Falmouth	(1903)	Unknown	22′ 2″ x 7′ 8″ x 4′ 5″
Harriet	Restronguet	1861	Wm. Ferris	28′ x 9′ 6″
*Helen Mary**	Mylor	1974	T. Heard	
Hope	Unknown	(1869)	Unknown	20′ 6″
*Ivy**	Calenick	1905	Wm. Brabyn	25′ 3″ x 8′ 2″ x 4′ 6″

NAME	PORT OF BUILDING	DATE	BUILDER	DIMENSIONS
Isadora	Unknown	(1874)	Unknown	30′
Ida	Rebuilt Restronguet	Unknown	Unknown	23′ 6″ x 7′ 4″ x 3′ 8″
*Irene** (ex Kayte)	Mylor	1978	T. Heard	28′
John Wesley	Unknown	(1895)	Unknown	29′
Joseph Eliza	Unknown	(1869)	Unknown	35′
Joseph Collins	Unknown	Unknown	Unknown	Unknown
Julie*	Mevagissey	Unknown	Unknown	20′ 9″ x 7′ 9″ x 3′ 9″
John	Mevagissey	(1874)	Unknown	28′
Jason	Mevagissey	(1896)	Unknown	30′
Josephine	Mevagissey	(1869)	Unknown	21′ 6″
Kestrel	Porthleven	1880	Unknown	30′ 5″ x 8′ 9″ x 4′ 8″
Kathleen (ex Alvina)	Mevagissey	1892	Unknown	26′ x 9′ 5″ x 5′
Kate	Unknown	(1869)	Unknown	23′
*Katherina**	Looe	1913	P. Ferris	23′ 8″ x 8′ 1″ x 3′ 6″
Lucky Boy* (ex Pathfinder) (ex Flying Angel)	Porthleven	1890	Oliver	25′ 5″ x 8′ x 4′ 8″
Liberty	Unknown	(1869)	Unknown	24′
Little Ellen	Unknown	(1869)	Unknown	24′
Lily*	Kings Lyn	1906	Worfolk	35′ 9″ x 9′ 9″ x 4′ 6″
*Leila**	Mylor	1963	M. Parsons	23′ 3″ x 7′ 6″ x 4′ 5″
Lorna Doone	Falmouth	(1920)	Unknown	27′ x 9′ x 6′
Lottie (ex Ella)	Unknown	Unknown	Unknown	26′ x 8′ 4″ x 5′ 2″
Lily of the Valley	Restronguet	1884	Unknown	28′ x 8′ 4″ x 6′ 3″
Lewis (ex Unexpected)	Unknown	(1871)	Unknown	25′ 5″ x 7′ 8″ x 4′ 4″
Lois	Unknown	(1922)	Unknown	24′ 5″ x 8′ 2″ x 3′ 8″
*Lena**	Mylor	1979	T. Heard	24′
Lilian* (ex Dorothy Louise)	Porthleven	1902	Unknown	30′ x 9′ 3″ x 4′ 6″
Lucy	Unknown	(1869)	Unknown	23′
Lucy	Unknown	(1869)	Unknown	22′
Lily of the North	Unknown	(1903)	Unknown	26′ 2″ x 8′ 8″ x 5′ 3″
*Molly**	Falmouth	1980	Cygnus Marine	28′
Magpie 2	Unknown	Unknown	Unknown	18′

NAME	PORT OF BUILDING	DATE	BUILDER	DIMENSIONS
Maggie Dixon (ex Saphire)	Unknown	Unknown	Unknown	Unknown
Maud (ex Lilian)*	Unknown	Unknown	Unknown	20′
Murre*	Mylor	1971	T. Heard	22′
Magdalena* (ex Lena)	Mevagissey	(1890)	Unknown	22′ 4″ 7′ 8″ x 4′ 6″
Melita*	Porthleven	(1903)	Unknown	23′ x 8′ x 3′ 7″
Mabel	Restronguet	1895	F. Hitchens	28′
Mayflower*	Porthleven	1903	Oliver	25′ x 8′ x 4′ 6″
Myrtle (ex Eva)	Porthleven	1899	Unknown	25′ 4″ x 8′ 8″ x 4′ 6″
Mildred*	Gorran Haven	1913	R.D. Pill	27′ 6″ x 5′ 6″
Maiden Nellie* (ex Boy Phillip)	Mousehole	1903	Unknown	26′ 6″ x 8′ 9″ x 4′ 6″
Muriel*	Unknown	(1945)	Unknown	24′ 3″ x 8′ 3″ x 4′ 6″
Moya*	Morecambe	1905	Arneside	34′ x 11′ x 3′ 6″
Morning Star	Unknown	(1840)	Unknown	33′ 6″ x 9′ 6″ x 5′ 3″
Melody	Unknown	(1869)	Unknown	45′
Maria	Unknown	(1869)	Unknown	27′
Mary Jane	Unknown	(1882)	Unknown	24′ 8″
Maria	Unknown	(1869)	Unknown	24′
*Mildred**	Calenick	1890	Wm. Brabyn	30′ x 10′ x 5′ 2″
Maid of the Mill	Unknown	(1875)	Unknown	35′
Mystery	Unknown	(1869)	Unknown	21′
Mary Ann	Unknown	(1859)	Unknown	25′
Majestic	Mevagissey	1896	Unknown	19′ 4″ x 6′ 3″ x 3′ 8″
Margaret	Mousehole	1903	Unknown	28′ x 8′ 7″ x 4′ 8″
Minnie	Unknown	(1879)	Unknown	Unknown
Nellie*	Looe	1904	P. Ferris	26′ x 8′ 9″ x 5′
Odessa	Unknown	(1871)	Unknown	24′
Ocean	Porthleven	(1903)	Unknown	31′ 4″ x 8′ 4″ x 4′ 8″
Olga Joan	Mevagissey	1929	Unknown	20′ x 7′ x 2′ 5″
Obediah (ex Chum)	Mevagissey	1936	Unknown	19′ x 6′ 4″ x 3′ 2″
Orlando*	Gorran Haven	1911	Unknown	26′ 4″ x 9′ 2″ x 5′ 2″
Onward	Mevagissey	1870	Unknown	28′ 5″ x 8′ 5″ x 4′
Olivia	Unknown	Unknown	Unknown	Unknown
*Orca**	Mylor	1979	T. Heard	28′
Pelican (ex Daily Mail)	Porthleven	1912	Unknown	25′ 4″ x 8′ 4″ x 3′ 1″
Pet*	Porthleven	(1895)	Unknown	32′ 5″ x 10′ 2″ x 5′
Prosperity	Unknown	(1869)	Unknown	28′

NAME	PORT OF BUILDING	DATE	BUILDER	DIMENSIONS
Pioneer	Unknown	(1869)	Unknown	60′
Pearl	Unknown	(1869)	Unknown	21′ 2″ x 6′ 8″ x 4′
P.C.	Falmouth	1886	Unknown	25′ 3″ x 6′ 6″ x 3′ 9″
Pearl	Unknown	(1869)	Unknown	25′
Pride of the Isle	Unknown	Unknown	Unknown	Unknown
Polly	Unknown	(1869)	Unknown	25′
Phantom*	Newquay	Unknown	Unknown	25′
P.C.W.	Unknown	(1869)	Unknown	28′
*Pandora**	Restronguet	(1903)	Unknown	22′ x 7′ x 3′ 4″
Petrel	Unknown	(1883)	Unknown	33′ 7″ x 8′ 5″ x 4′ 8″
Queen of the Bay	Calenick	1892	Wm. Brabyn	23′ 7″ x 7′ 4″ 4′ 6″
*Queen of the May** (ex Andromeda)	St Just	1906	Pascoe	30′ 6″ x 9′ 8″ x 5′ 3″
Queen of the Fall	Unknown	(1869)	Unknown	21′
Rover	Falmouth	(1869)	Unknown	26′ x 7′ 8″ x 5′
Rob Roy	Unknown	(1869)	Unknown	45′
Ripple	Unknown	1866	Unknown	22′ 8″ x 7′ 8″ x 3′ 2
Rambler	Unknown	(1869)	Unknown	23′
*Rita** (ex Melorus)	Mylor	1969	T. Heard	28′
Rebecca	Unknown	(1892)	Unknown	26′
*Result**	Mylor	1961	T. Heard	29′ 9″ x 7′ 6″ x 4′ 9″
Royal Oak*	St. Ives	1902	Paynter & Co	23′ 6″ x 8′ x 4′
Rippling Wave (ex Shagreen)	St Ives	1902	Unknown	25′ 7″ x 8′ 5″ x 4′ 9″
*Serica** (ex E.R.C.)	Looe	1912	P. Ferris	24′ 4″ x 8′ 2″ x 3′ 7″
*Softwing**	Restronguet	1900	F. Hitchens	24′ x 8′ x 4′ 3″
Swallow*	Unknown	(1892)	Unknown	27′ 6″ x 8′ x 3′ 6″
*Shadow**	Restronguet*	1870	F. Hitchens	27′ x 8′ 6″ x 4′ 6″
*St Melorus**	Mylor	1968	T. Heard	28′
Shag	Rebuilt Restronguet	1880	Wm. Ferris	26′
Six Brothers (1)	Feock	1890	Wm. Ferris	28′ x 8′ 8″ x 4′ 5″
Six Brothers (2)*	Plymouth	1969	Skentleberry	28′ 5′ x 9′ 5″ x 6′ 1″
Skandinvia	Falmouth	1899	Unknown	21′ 5″ x 8′ 1″ x 4′ 5″
Seabird*	Gerrans	1890	Unknown	20′ 8″ x 6′ 9″ x 3′ 6″
Sonny South*	Newquay	Unknown	Unknown	25′ x 8′ 4″ x 4′
Shamrock*	Mevagissey	1908	Unknown	24′ 2″ x 8′ 4″ x 4′ 9″

NAME	PORT OF BUILDING	DATE	BUILDER	DIMENSIONS
Seacroft (ex 'Dawson')	Porthleven	(1945)	Unknown	25′
Swift	Unknown	(1883)	Unknown	25′
Stella*	Porthleven	1909	J. Bowden	25′ x 8′ 6″ x 4′ 3″
Silent	Mevagissey	1888	J. Bowden	38′ x 10′ x 5′
Sunbeam (ex Phyliss A)	Looe	1912	Unknown	25′ x 8′ 4″ x ?
Serene	Porthleven	1905	Unknown	32′ 9″ x 10′ 4″ x 4′ 2″
Sprat	Unknown	(1896)	Unknown	22′ 3″ x 7′ x 3′ 7″
Snowdrop	Unknown	(1913)	Unknown	18′ x 6′ 1″ x 2′ 9″
Susan Jane	Unknown	Unknown	Unknown	Unknown
Shooting Star	Mevagissey	Unknown	Unknown	28′ – 30′
Seagull*	Itchen Ferry	Unknown	Unknown	21′ 1″ x 8′ 6″ x 3′ 6″
Sonny	Porthleven	(1896)	Unknown	26′ x 5′ x 6′
Seacroft	Porthleven	Unknown	Unknown	25′
Twilight*	Looe	1903	Unknown	36′ 8″ x 11′ 3″ x 6′ 2″
Torment	Unknown	(1918)	Unknown	16′ – 18′
Three Brothers (ex Petrel) (ex 'Lady Beatrice')	Penryn	1928	Unknown	22′ 5″ x 7′ 8″ x 4′
Twins	Unknown	(1869)	Unknown	22′
Thelma	Restronguet?	Unknown	Wm. Ferris?	18′
Teaser	Unknown	(1869)	Unknown	22′ 3″ x 7′ 4″ x 3′ 8″
Thetis	Unknown	(1869)	Unknown	22′
Teazer	Unknown	(1877)	Unknown	27′
Two Brothers	Unknown	(1870)	Unknown	50′
Tern*	Looe	1920	P. Ferris	26′ 9″ x 9′ 8″ x 5′ 6″
Two Sisters	Unknown	(1895)	Unknown	28′
*Three Sisters**	Mylor	1973	T. Heard	28′
Unique	Unknown	Unknown	Unknown	Unknown
*Veruna**	Mylor	1978	T. Heard	24′
Vera* (ex White Heather)	Porthleven	Unknown	Unknown	32′ 3″ x 9′ 9″ x 5′ 3″
Vic	Unknown	(1905)	Unknown	27′ 4″ x 9′ 1″ x 5′ 3″
Vanguard	Unknown	Unknown	Unknown	Unknown
Vivian* (ex Bessie) (ex Winifred)	Looe	(1919)	Unknown	21′ 6″ x 7′ x 3′ 8″

NAME	PORT OF BUILDING	DATE	BUILDER	DIMENSIONS
*Victory** (ex Royal Oak)	Restronguet	1882	F. Hitchens	27′ 2″ x 8′ 7″ x 5′ 3″
Victoria	Unknown	(1869)	Unknown	22′
Winkle Diver* (ex Tosser)	Mevagissey	1934	Unknown	21′ x 7′ 3″ x 4′
Water Hen	Mevagissey	(1903)	Unknown	24′ 9″ x 8′ 4″ x 5′ 4″
White Heather?*	St Ives	1905	Unknown	24′ 8″ x 8′ 8″ x 3′ 8″
Wild Cat	Falmouth	(1903)	Unknown	30′ 3″ x 8′ 2″ 5′ 6″
William	Unknown	(1869)	Unknown	21′
William Green	St Mawes	1864	Peters?	29′ 9″
W.H.B.	Unknown	(1895)	Unknown	24′
*Winnie**	Calenick	1898	Wm. Brabyn	28′ 3″ x 8′ 6″ x 5′ 9″
William & John	Unknown	(1869)	Unknown	35′
Willie	Unknown	(1895)	Unknown	30′
*Zephyr**	Restronguet	1953	N. Ferris	23′ x 7′ 8″ x 3′ 9″
Ziguener	Germany	(1840)	Unknown	29′ x 8′ 4″ x 5′
Zephyr	St Ives	1906	Unknown	26′ 4″ x 9′ 2″ x 3′ 6″
Zona*	Mevagissey	Unknown	Unknown	30′ 7″ x 10′ x 5′

It can be seen that of the boats built for the job, the greatest number came from Restronguet, Mylor, Calenick and Pill respectively, and is only to be expected that most should have been built at Restronguet as many of the dredgermen lived around the creek. Even though more boats worked from Pill Creek, it does not provide as much in the way of good boat-building facilities as Restronguet Creek, as there is little space above the high water mark, due to the steep cliff sides.

HARRIET

It was here in 1862 that the first of William Ferris' Falmouth Working Boats, 'Harriet', was built for John Penna of Pill. By 1886 she was still at Pill in the Ferris family though owned by James Ferris who was probably one of William's eleven sons. James sold her as a yacht in 1934 and the next time we hear of her she was laid up at St. Mawes. In 1964 she was bought by D. Martin who raced her during that summer, however, by 1968 she was in a pretty bad way, and when a leg broke and she fell on her side, Terry Heard decided to end her days with a bulldozer. Part of her still lives on though, as her iron keel was put on the boat that Terry Heard was then building, 'St. Melorus'.

'Harriet'. To quote Basil Greenhil in his book 'Merchant Schooners' again, "a vessel of singular beauty of line and a great local reputation for speed."

'Harriet' was a perfect example of the type 28ft long very beamy (9ft 6ins) and as much draught as she could easily use dredging. She was built without caulking as were all 'Foreman' Ferris' Working Boats, the planks being planed to fit. This method can have drawbacks, as was discovered on the day that she was launched, for a gap has to be left to allow the wood to swell. As a result it meant that she leaked quite a lot, causing great consternation when it was realised that a child had been launched with the boat, which was now sinking fast. However, all ended well and both child and boat survived the launch!

FLORENCE

Next was the 'Florence', built thirty-three years later in 1895 at Ponsmain, which is on the opposite side of the creek to where 'Harriet' was built. The cost of building was £65, complete with a dingy and spars: 'Foreman' Ferris built her for Richard 'Budes' Ferris and she stayed in the Ferris family until 1955, when she was sold to T. Burley. He had her practically rebuilt by Bart Moore and dredged in her himself the following season. She is particularly well known for her recent racing career which commenced about 1965 when T. Burley lent her to J. Sawle of St. Mawes. Previously a top Sunbeam helmsman, with the help of his crew, he restored her to pristine condition (far better than most yachts) and crowded a mass of new canvas on her (half as much again as most other boats). In fact she looked more like one of the pre war J. class yachts than a Falmouth Working Boat and even though she has since returned to dredging, she was the fore-runner of the trend of over canvassed 'yachts' in the working boat class. Her most successful year was 1970 when she took nine first prizes, six second prizes and three thirds. As the only boat sailing from St. Mawes she was taken to heart by the 'Turkeylanders' (the nickname for the St. Mawes villagers), who supported her to the extent of raising money for new sails at coffee mornings. She was, however, always thought to be something of a white elephant as a dredging boat because of her exceptional draught a foot more than most, a charge which was not justified as she has worked successfuly for many years. She was modelled on a boat called 'Zephyr' which was 18ft long and 4ft 5ins draught and built in 1892 by 'Foreman' Ferris for a Dr. Harding. Jack Ferris (who owned the Pandora Inn) one of 'Foremans' sons and who owned 'Florence' in 1925 remembered 'Zephyr' being built, as he used to help his father build her when he came home from school. Both 'Florence' and 'Zephyr' were built on the same beach. She left the Ferris family for the first time in 1978 when she was sold to John Jackson (who had now returned from his trip to the West Indies in 'St. Melorus'). He sold her in 1979 to Alun Davies who returned her to dredging and who still owns her. Her nickname in the 'old days' used to be the 'Silent Sailor', a reference to the fact she leaves such a clean wake, and it was said that Wm. Ferris put a concave curve in her garboards to help her 'grip the water'. They used to say that she was shaped like a plough "to hold the water and throw the water"!

SIX BROTHERS

The last of the Ferris boats, but by no means the last important, was the ill-fated 28ft 'Six Brothers' which tragically came to grief on Greatwood Quay during a S.E. gale on Thursday 21st November 1968.

The 'old' 'Six Brothers'.

Foreman built this last working boat in 1898 at Ponsmain, where he built 'Florence' and 'Zephyr'. She was a compromise in shape between 'Florence' and 'Harriet' and perhaps the fastest of the three. The first owner was Captain Jimmy Lewarne of Helford Passage (an ex schooner man) who was one of six sons, hence the name. He had her built for fishing, though she probably dredged too. He sold her in 1935 to Mrs. Hext of Trebah House, Helford who used her for pleasure in the Manacles area. Next was Richard Winfrey in 1942, a farmer in Frenchmans Creek then a Mr. Forsythe until finally in 1953 she was bought by her present owner, G. Vinnicombe of Mylor. When he found her she was a derelict, having broken adrift and gone on the rocks, so he towed her to B. Moore's yard with his fishing boat 'Boy Willie', where she was 'done up'. Her most successful racing year was 1965 (the first year 'Florence' was 'done up'), when she had eight firsts, three seconds and four thirds. Had she had the money and attention lavished on her as did the 'Florence', who knows how she would have done.

George used her for dredging up until the day she was sadly wrecked in 1968. All that was saved was the transom (which had her name on it, her keel and frames. However, with the help of a half model, photos and the remains, P. Dalton managed to make a plan of her and a new Six Brothers was commissioned and built by Skentleberry of Plymouth in 1969. She appears to be the same as the old Six Brothers in most respects, however, due to a builders error her transom was put four inches too high, giving her a higher transom than the original, and a raised sheerline aft.

The 'new' 'Six Brothers' The difference in the sheerline clearly visible.

SHADOW

'Shadow'

The rivals to the splendid boats of 'Foremans' were Frank Hitchens 'Shadow', 'Victory', 'Mabel', 'Evelyn' and 'Softwings' and that also was the order in which he built them.

His first boat 'Shadow' was built in 1868 or 1870 (probably 1870) at Point for Thomas & William Hitchens. She is 27′ x 8′6″ x 4′6″ and the narrowest of all his boats, each successive boat being fuller in the bilge. She has always had a reputation as a fast boat and for many years held the record for the Fowey-Falmouth passage race with a time of two hours forty-nine minutes (an average speed of seven knots). She stayed in the Hitchens family until 1935 when she was sold as a yacht to Mr. A.C. Visick. He kept her until 1944 then sold to Thos. Trebilcock from whom she passed to her present owner, F.F. Vinnicombe an oyster merchant of Mylor who dredges in her himself.

MABEL

His next boat 'Mabel' built in 1895 was about 27ft. and built for Edward Green. Little is known about this boat as she was sold to Germany as a yacht, where she probably still remains. She was lucky to get there, for as she was coming back from fishing in Falmouth Bay one night, the helmsman fell asleep and she went aground on St. Anthony Point. She stayed in Falmouth until 1927/28 at least, and during her last years here was owned by 'Dumper' Morrisson who used her taking stores to the 'Elevator' off St. Just. She was rigged with a short mast on which was set only a 'leg of mutton' steadying sail, and had a big engine installed. Both she and 'Orlando' were rigged like this and acted as Quay Punts on the 'Elevator'.

The only known photograph of Hitchen's 'Mabel' shows what a fine boat she was, seen here at Loe Beach Regatta, date unknown. Her wake is evidence of her speed.

The original 'Victory' leads a fleet in front of the steam yacht 'Rover'. She carries a mizzen as most Quay Punts did.

The present 'Victory' (ex 'Royal Oak') racing off Fowey.

After 'Shadow' was 'Victory', who started life as the 'Royal Oak' in 1884. She is 27′ x 8′7″ x 5′3″ and is the most well known of Hitchens boats. Her present owner Arthur 'Toby' West has owned her and raced her for many years at Falmouth and both have always been the firm favourites of the Falmouth waterfront. She is similar to 'Shadow' but slightly deeper and broader at the waterline. She was first owned by Peter Jacobson (Restronguet) who sold her to H.K. Neale in 1925, who kept her as a yacht. She came into the West family when 'Victory' (a quay punt) owned by Toby's father, was wrecked on Well Beach about 1925. The locals had a 'round robin' and bought him the 'Royal Oak' for about £80 or £90, and he changed her name to Victory. During the time P. Jacobson owned her and fished mackerel in her she had a reputation as being very slow, but the West family soon altered that when they started racing her. They raced her under Gunter and Bermudan rig, but changed back to gaff in 1956 and has stayed gaff rigged ever since.

EVELYN

In 1898 Hitchens built 'Evelyn' the fastest of all his boats and debatably the fastest Working Boat ever built. She too was built at Yard Point, for Thos. & James Hitchens (Feock) and registered for nets, lines and crabbing, and they kept her until 1905 when she was sold to John Billing of Flushing. She was named 'Evelyn' after James Hitchens daughter. It was her next owners, R.A. & W. Billing (Flushing) who bought her in 1917 who had the iron keel put on her, but they only used her for trawling and crabbing. When she was built she was intended to be the fastest of the fleet, but though she carried a lot of canvas, she never could live up to her expectations until Jackits at Falmouth put on the iron ballast keel for the Billings Bros. After that they always swore she was the best of the lot and her recent racing career, under her owner Kenny Cork will bear witness to that. The Billings Bros. sold her in 1935 to an elderly couple at Helford when she stopped fishing and for a time had her name changed to 'Curlew' probably by next owner Dr. Hollington (Sales Manager at Boat Construction) of Coombe Creek, where she remained until bought in Sept. 1958 by A. Mathews a Penzance sailmaker, who changed her name back to 'Evelyn'. Finally in 1965 she returned to Flushing where she was 'reconverted' (coachroof removed etc.) and put back to work. In 1988 she was sold to Bernie Trenoweth who has largely rebuilt her prior to taking her out racing again, as she has not raced for the last few years.

'Evelyn', regarded by many dredgermen as the ultimate hull shape for the job. Full bilges allow a reduction in draught, a major consideration when dredging over the shallow banks.

This is a copy of the advert listing the inventory with "Curlew" when she was put up for sale by Dr Hollington in 1958

GAFF CUTTER "CURLEW" Originally 'EVELYN'

Length. 28' Beam 9' Draft 4' 6"

Engine 4 cyl. Panther Marine 13/15 hp. Petrol/TVO giving an approx. speed of 6 knots under power. A proper marine engine not a conversion, with a bilge pump.

Equipment. 1 25lb CQR anchor 1 20lb fishermans anchor (new 1953)
25 fathom. 3/8 galv. short link chain (new 1953)
90' 2" Coir warp.
9' Boat hook with brass head.
1 30" Lifebuoy.
1 2gall. Minimax fire extinguisher.
1 3gall. varnished oak water breaker on stand.
1 Compass.
2 Oil cabin lamps.
1 set navigation lamps, copper.
1 " " " shades on shrouds.
1 silent primus on gimballs.
1 Baby Blake toilet.
1 Canvas bucket.
1 Simpson Lawrence bilge pump.

Fuel Tanks. TVO 7galls (copper) Petrol 2½ galls (copper)

Construction. Pitch pine on oak all full length planks, no butt ends. Bobstay chain ¼" galv. short link set up by Handy Billy with galv. iron blocks.

Stem fastening through double bolted
All rigging blocks yacht quality elm and varnished and internal bound.

Mast: Pitch pine 28' 6" from truck to deck 32' from truck to keelson.

Deck. laid deck 1½" timber.

Accomodation. Fo'c'le contains self stowing chain locker under naval pipe, shelf rack, one bunk complete with cushions, Baby Blake toilet, silent primus on gimballs in cooking fixtures, sail shelf, fore hatch, 3 brass opening port lights. 1 fixed port. light brass rim in fore hatch. One galv'd. mushroom ventilator over sail shelf. A bulkhead and sliding door devides focs'l from main cabin which contains one bunk complete with cushions, One china locker. One supply locker under both setee & bunk, also one metal locker. Two brass opening port lights and six fixed port lights with brass rims

Rigging.

Shrouds. 1½" galv. plough steel wire set up by ½" dia. shoed galv. rigging screws.

Forestay as above.

Topmast stays. 1" Galv. plough steel wire set up

by 5/16" thread galv. rigging screws.
Backstays. 3/4" galv. plough steel.
Jib Sheets. 1/2" American cotton rope. Other cordage sisal. Peak halyar, throat halyard & main sheet renewed every other season.
Working sails. Jib, Mains'l, Stays'l all waterproofed & dyed birch red. Large genoa jib in white American cotton canvas.
Under Construction. One tops'l in white American canvas and one 13' tops'l yard.

The 'Evelyn' was built by Hitchins of Restronguet about 1910. She did four years as a fishing boat and was then sold to a gentleman at Helford who had her converted to a Yacht. She was later sold to another owner from whom the present owner bought her.

As far as is known only three boats were built to this design, the 'Evelyn', the 'Shadow' and one other. The 'Evelyn' never raced but the 'Shadow' won nearly all her races. Hitchins had a first class reputation as a boat builder. The four owners of the 'Evelyn' have always done their best by her and she is recognised as one of the best examples of her type in the Westcountry.

SOFTWINGS

The last boat 'Softwings' was built for Edward Green (as was 'Mabel') to replace 'Mabel' which he sold in 1899. She was the smallest of all his boats, being only 24′3″ x 8′ x 4′3″. She is more like 'Evelyn' in shape than any other, being very full in the bilge.

In 1902 she was sold to W.E. 'Pyman' Ellis (Pill). In 1958 Robin Vinnicombe bought her from 'Pyman' Ellis for dredging and used her for shark fishing in the summer; about 1970 she was purchased by the National Maritime Trust who have had her restored to first class order and got her sailing again in her home waters.

QUEEN OF THE BAY

Further up the river at Calenick Steven Brabyn was equally busy. His first boat was 'Queen of the Bay' which he built in 1892. She was one of the smaller boats, 23′7″ x 7′4″ x 4′6″. It is not known who she was built for, but the first record of ownership shows Sam Furse of Flushing in

'Queen of the Bay' watching the start of the Scratch Boats 'Harriet' leads 'Ziguener' and 'Morning Star'.

1904 who had the misfortune to drown in her. She passed on to James Ellis of Feock in 1906 who sold her in 1932, when she went up country and has not been heard of since. She had a reputation as a good light weather boat, though was a bit tender in a breeze.

ECHO

In 1898 his next boat 'Echo' was built for William 'Barky' Pearce of Point. She was about 23′. He kept her until 1925 when she was sold to William Pearce of Devoran. She was his pride and joy and he kept her more like a yacht than a working boat, and even put a fidded topmast on her. This was very unusual on a working boat and she was obviously more of a plaything than a tool. The last known owner was Herbert Monk who sold her to somebody in Mounts Bay. It was later heard that she was taken on passage up channel, foundered and was lost.

WINNIE

1898 must have been a busy year for Brabyn as he was to complete one more boat, 28′ 'Winnie' was built for E.A. Merrifield who in 1944 passed her on to his son Henry Merrifield (owner of 'Endeavour') and finally in 1969 left the family for the first time when she was sold to S.J. Keates to be used exclusively for racing. A big powerful boat,

'Winnie', a powerful boat that can carry her sail, here has problems with the bobstay. The result of photo page 112.

28′4″ x 9′3″ x 4′8″, she was used for trawling during the summer months around 1939. In 1920 she had a 9h.p. Kelvin put in her but later had a

13h.p. Kelvin put alongside to give her more power for trawling. She was built of pitchpine on oak, bought from two timberyards. Joe Keates sold her to Karl Martin who kept her for a few years before selling her to her present owner, Arthur Williams, in 1988.

MILDRED

When Brabyn completed'Winnie' his next boat was 'Mildred'. She is not to be confused with the 'Mildred' built at Gorran Haven and owned by David Stonehouse. This 'Mildred' left Falmouth many years ago, probably in the 30's as that is when she was converted to a yacht. She was built as an almost identical boat to 'Winnie', the same shape but slightly longer, beamier and deeper (29′6″ x 10′2″ x 5′2″). She was built for Wm. Thomas and Earnest Gray (Restronguet) and named after his sister. They kept her until 1906 when she was sold to Falmouth. There is no record of her again until 1957 when she turned up at Shoreham owned by a Mr. Northeat. He kept her there until 1965 when she was bought by Mr. Lingard of Worthing, where she was moored until 1973 when he sold her to a Frenchman who intended taking her back to France. She is now owned by A. Köpne in Keil, Germany and renamed 'Turf'.

BEATRICE

In 1903 Brabyn built 'Beatrice' for Stephen Brabyn (his brother) who lived at Point. She is 24′6″x7′9″x5′ and has a cutaway forefoot as has Brabyn's last boat 'Ivy' which is very similar to 'Beatrice' though slightly shallower aft. Brabyn kept her until 1924 when he sold her to Samuel Furse (Flushing). He worked her but laid her up between the years 1941

'Beatrice', the perfect single handed boat dredging on East Bank with 'Evelyn' and 'Six Brothers' behind her.

to 1944 and in 1949 she came under the ownership of Leonard Furse (Flushing). She stayed with him until 1956 and did not come back fishing again until 1967 when John Walls (Mylor) bought her from W.H. Rowe (Falmouth) who had kept her as a yacht. John Walls kept her until 1973 when he sold her to Dicky Rees (Penryn) who used her for dredging in the winter and mackereling and longlining in the summer.

IVY

Brabyn built his last boat 'Ivy' in 1905 for Thos. & James Hitchens who had just sold their previous boat 'Evelyn' as she was too slow and they left her on a mooring for the last two years they had her for that reason! She is 25′ x 8′1″ x 4′5″, and after Hitchens sold her she went through quite a few hands. In 1913 she was sold to Thos. James (the owner of the Pandora Inn at Restronguet), in 1930 to Margaret Symons

'Royal Oak', 'Ivy' and 'Serica' at Pill Creek.

(again of the Pandora Inn), in 1951 to R.G. Symons (Feock), 1956 R.T. Symons (Feock), 1958 to George Newman (Feock) with whom she stopped fishing in 1961. She was owned by Richard Berey in 1964, some time after this Dick Berry bought her and said that at this time she had a counter stern, which was cut off by her next owner Aubrey Ferris of Pill.

SERICA

The first boat that Peter Ferris built at Looe and sent down here to dredge oysters was 'Serica', though at that time named 'ERC'. Her name was obtained from the initials of the two owners – Edward Robert Ellis, his brother Cecil providing the C. In 1958 she was purchased by her present owner Paul Williams who changed her name to 'Serica.' She is typical of all Peter Ferris's boats in that she has a very cutaway

The extreme underwater lines of Paul William's 'Serica'.

forefoot. Her dimensions are length 24′9″, beam 8′2″ and draught 4′9″. If she has a claim to fame it is surely that she is the first Working Boat to be equipped with Hood sails and they certainly seem to have done the trick, for during the 1974 racing season she won no fewer than 8 cups out of 10 races.

KATHERINA

1913 saw the arrival of the second Ferris boat from Looe, 'Katherina'. She was built for James Ferris (Penpol) and was the same size and shape as the other boats he built, but at 23′8″ x 8′1″ x 3′6″ just the smallest. James Ferris sold her to Roger Hemming (Penpol) in 1920, who kept her until 1932 and sold her to George Blamey (Kea). She later returned to the Ferris family for a time when Richard Ferris bought her in 1940, James Ferris (Feock) in 1945 and in 1968 was sold to Ray Frost of Feock,

John Moore stows the staysail on 'Katherina' at the end of a days dredging before he motors home to Mylor.

(owner of 'Stella'), and his son Colin worked her until 1972 when she was sold to Derek Veevers of Trevissome, who dredged in her for a year before selling her to John Moore who also dredged with her for a time. She passed on from John Moore to her present owner Jonothan Bailey in 1974 who dredges her in the winter. His brother Tim Bailey used to race her very successfully in the summer.

FIVE SISTERS

The next boat to arrive here from Looe was '5 Sisters' in April 1914; she was built for Thos. Ferris (Feock) and has caused confusion in many quarters ever since. Though her correct name is 'Five Sisters', Thos Ferris's previous boat (an old 'tucking' boat from the days of seine netting for pilchard), was called 'Alfred' but nicknamed 'Dolly', so when the new boat arrived it promptly became the 'New Dolly'and consequently some people have always known her as 'Dolly', others as 'Five Sisters' and some as 'New Dolly! In 1921 she passed to Thos. Ephraim Ferris (Kea) the son of the man who had her built, who kept her until 1940 and then sold her to John Boyden (Mylor). He kept her only one year, selling her to William Trebilcock (Carnon Downs) and he in turn sold her to John Cock of Restronguet in 1954. John Cock did

Barry Prynn at work in the 'Five Sisters'.

a lot of work on her, putting 4 planks in each topside, new stern, deck and spars, and raising her topsides one plank. Whilst he owned her she broke adrift Christmas Eve and went ashore on rocks by Restronguet Pond. The keel dug in six or nine inches and they had to remove her ballast in order to float her off. She was sold to her present owner Barry Prin (Mylor) in 1972 who uses her for dredging still. Similar to 'Serica' she has a cutaway forefoot but is a bit deeper having a 5′2″ draught on 25′ x 8′6″.

TERN

The last Ferris boat was the 26′9″ 'Tern'. He built her in 1920 for James Ferris and the design was probably largely James Ferris's as she is quite different from his other boats, though she has curved stem and cutaway forefoot. Her draught is 5′6″ and beam 9′8″. She was very strongly constructed her frames being grown 2″ x 1½″ double on 16″ centres and in 1972 was in first class condition. In 1933 she was owned by Russel Myners, Gordon Martin and Capt. 'Skip' Wallace – possibly Hollington of 'Evelyn' fame bought her next and lived on her at Pill until about 1956 when a Mr. R. Clare bought her and took her to Salcombe. He lived on her there for 4 or 5 years until his family outgrew her and in 1961 a sailmaker John McKillop bought her. At first he used her for charter work though latterly he acquired two partners and used her for cruising and racing. He did a lot of work on her (6 or 8 new planks, new transom, short stringer by chain plates) though he did not put on the coach roof she now sports. In 1971 she won the Solent Old Gaffers Race, though when she came and raced at Falmouth during Falmouth week shortly after she had little success against the Falmouth boats, finishing last in every race! In 1973 she was sold to an American who plans to sail her to the States. She has a very heavy keel for a boat of her size, two tons instead of the more usual 1½ tons.

Tern

RESULT

In 1960 Terry Heard started work on his first working boat, she was to be the 26′9″ 'Result' (named after the schooner) and he worked on her in his spare time during the winter of 1960-61 and she was launched in 1961. She was designed as a compromise between 'Florence' and 'Shadow' as can clearly be seen from her side profile, however, she did not manage to retain their bilge or beam. Being long and narrow (7′6″ beam) she sails well off the wind when she is upright, but stops dead on the wind, as without sufficient beam or bilge to hold her up she just lays over flat. In an attempt to correct this failing Terry Heard cut right through her horizontally on the waterline and leaving only 4 or 5 planks in the bottom and 4 or 5 in the top, tried to force her open a bit more, and put in a stealer. This was done in about 1965 and two years later she was sold to Henry Pickivance when she started her dredging career. He worked her for 4 yrs. and in 1970 she was sold to Bob Harvey George

'Result' in the garage before Terry Heard modified her.

who worked her on and off, then she was purchased by Alun Davies in 1972 who worked her and raced her until 1974 when she was sold to J. Bellamy and taken by lorry to Bristol.

CHRYSTAL SPRING

Terry's next boat was the 20 'Chrystal Spring' in 1966. Built on similar lines to a crabber i.e. very flat floors, she was worked initially by Dave Cain who sold her after a few years and she is now in the Exmouth area. Terry named her after a schooner wrecked off Bude.

Tony Murray bending in a plank on 'Chrystal Spring'.

ST MELORUS

In July 1968 came the boat for which Terry Heard is best known. Designed by Percy Dalton it became the mould for the first fibreglass working boat to be built (also by Terry Heard). She was called 'St.

'St Melorus' racing in the sunshine of Antigua after John Jackson sold her in the West Indies at the end of his trip.

Melorus' after the patron saint of Mylor, though locally known as the 'fire engine', due to her colour scheme, bright red. She was worked during her first winter by George Vinnicombe who had just suffered the loss of the 'Six Brothers' and he raced her for her first two summers. However in the winter of 1969 she was sold to John Jackson who converted her to sail the Atlantic. He decked her in, gave her a coachroof and reduced her sail plan in order that she should meet the demands of the voyage.

Before she went Terry Heard cleaned her hull up and took a mould and built himself another one in GRP to replace her.

MURRE

The last wooden working boat Terry Heard built was the 22′ 'Murre'. This he bought nearly complete from a Mr. Prince and he completed her himself in 1971. She was designed by local designer Percy Dalton (who designed the 'St. Melorus' and helped on the new 'Six Brothers') Terry did not keep her for long and she was purchased next by David

'Murre' bouncing through a bit of sea.

Stonehouse who raced her for two seasons. She was sold to Peter Divers in September 1974 who dredged in her for a time. Her present whereabouts are unknown. These boats were all built at Terry Heard's yard at Tregatreath, Mylor at the head of Mylor Creek.

GAZELLE

The only other builder to build more than one boat was Norman Ferris, who lived at Point Quay. His first boat 'Gazelle' is 22′ and was built in 1940 for his brother W.J. Ferris. He sold her to a Mr. Dookson who in turn sold her to Charley Trebilcock. She finally arrived back in the family as she was sold to Norman 'Aubrey' Ferris (Norman's son) who still owns her. She first started dredging under the ownership of C. Trebilcock and for a while was hired out by 'Aubrey' Ferris to Gerald Pill who dredged her for a few seasons.

Not a typical Ferris design, 'Gazelle' has a generous beam, full bilge and shallow draught.

ZEPHYR

His other boat, the 23′ 'Zephyr', was also built at Devoran. An admirer of his grandfather Norman Ferris put a bit of all his famous ancestor's three boats into Zephyr, and though smaller, is a classic Ferris boat in style. She was built in 1960 and remained in the hands of Norman Ferris until his death in October 1974, and is now owned by 'Aubrey' Ferris.

'Zephyr' on launch day in 1958. Her builder and grandson of 'Foreman' Ferris, Norman Ferris, is second from the right.

SHAG

Yet another member of this famous Ferris family who built a Falmouth working boat was John 'Jack' Ferris at Restronguet. He built the 30' 'Shag' at Restronguet about 1880 apparently, though as she was a clinker built boat and all the rest of the local boats were carvel built, it may be that he did not actually build her but possibly rebuilt her. One or two boats have been credited with one particular builder, but research has proven them to have been merely rebuilt by him. In the case of the 'Shag', which was originally a counter stern boat, it may have been that she was originally a yacht and 'Jack' Ferris converted her to a more traditional working boat hull. Anyway, this is pure conjecture, but we do know that she passed into the hands of the Vinnicombe family where she stayed until the end of her days, when she was wrecked on St. Just during a gale. She was owned for many years by 'Father Dicky' Vinnicombe and Frank Vinnicombe. One day whilst sailing up Falmouth Harbour she hit a gust off Penryn River breaking her tiller. Unfortunately she sailed straight into an American barque moored ahead of her, hitting it square on with her bowsprit. The bowsprit was pushed straight back and hit the mast and knocked it down, which shook out all the caulking in her bottom and in seconds she was sinking fast. Luckily a tug happened to be close by, cast her a line and towed her to safety.

LEILA

Though many other boats may have been built locally for dredging we can only trace the builders of three of them with any degree of certainty. When M. Parsons first boat the 18′ 'Thelma' was wrecked on Greatwood Quay the dredgermen had a 'round robin' and provided him with sufficient cash to buy the materials and in 1962 he built a boat himself. He named the boat 'Leila' after his young daughter and she was completed in 1963.

1963 and 'Leila' on her way to the water for the first time.

The sails are bent on but 'Leila' is still a bit light of ballast up forward.

Though identical in rig to the rest of Falmouth working boats her hull shape was a slight departure from tradition. The lines were taken from a Itchen Ferry cutter called 'Centipede'. She is 23′3″ x 7′6″ x 4′5″ and differs from most boats in that she is very beamy for her length, very beamy on the waterline and has a high flat bilge. She has been a great success racing and working and Mike's only regret is that he did not build her a few feet longer. She has 'tumble home' on her topsides (as has the Gorran Haven 'Mildred') which is quite an unusual feature in working boats though many modern racing yachts have it.

ALF SMITHERS

One recent wooden boat to be built is the 26′ 'Alf Smithers' designed by Percy Dalton and built by boatbuilders Alf Smithers and Peter Mays. Work first started in 1968 when Alf and Pete were working together at their yard in Penryn. After Alf died about 1970 Peter moved the boatyard and boat to Cowlands Creek where he completed

Still awaiting completion, 'Alf Smithers' sits on the beach in Cowlands Creek.

and launched the hull in 1972 and Pete Mays started dredging with her in 1976. Originally designed as a 25' boat they decided to stretch her to 26′. She is similar to the first working boat Percy Dalton designed, the 'St. Melorus', though she is of course shorter and possibly a bit fuller on the waterline than the 'St.Melorus'.

It can now be seen that though a few Falmouth working boats were built for the job, the majority came from other fisheries, some close at hand others further afield.

There were other boats built at the creeks already mentioned (Pill, Restronguet, Calenick etc.) which were built for fishing.

For instance in Restronguet were built 'Emma', 'Chance', 'Ida' and 'Lily of the Valley'.

EMMA

'Emma' was built there possibly by Ferris and Hitchens for J. Blight of Flushing before 1869 but was more of a quay punt. She was dandy rig and very deep on the heel and though she dredged from time to time she was far from ideal. She was eventually sold to the Scilly Isles as a flower boat.

CHANCE

One of William Ferris's sons, also called William, built the 'Chance' in the creek about the turn of the century. He also built 'Vanguard'. 'Chance' was sold to go for a yacht in 1927.

IDA

Definitely not built as a fishing boat, 'Ida' adapted very easily to the role, though here seems to have reverted to her yachting days.

'Ida' was one of the few to reverse the process as she started life as an unregistered yacht but had her mizzen removed and was converted for dredging. She was 24′ with a counter stern and was broken up in 1950. In 1884 George Vinnicombe (later to own 'Ziguener') and his brother Richard had 'Lily of the Valley' built. According to the fishing register of 1884 she was built only for longlining and it was not until 1903 under the ownership of P. Vinnicombe that she started oyster dredging (as well as trawling, lining and netting). Unfortunately she was wrecked in 1929.

Of course we can never be sure about what job a boat was employed in unless somebody can remember, it may have been built for oyster dredging but that fact was omitted from the fish register for some reason, and in the absence of any accurate information a guess can only be made as to what exactly happened.

DEERFOOT

The only other boats known to have been built at Restronguet were the 'Deerfoot' and 'Pandora'. 'Deerfoot' was a 28′ boat built before 1869 at Pill for Stephen Ayma. In 1896 she was sold as a yacht but returned as a fishing boat to Falmouth in 1913 until 1920 when she was struck off the register (No Longer Fishing); on her return to Falmouth she was renamed 'Xerifa'.

PANDORA

The 'Pandora' was built prior to 1903 possibly for Joseph Paul of Flushing for the princely sum of 4 kegs of beer. She ceased fishing in 1932 and now lies at Point as a yacht.

'Pandora' at Point.

Porthleven Boats

Fortunately the smaller fishing boats used at Porthleven were suitable for conversion to oyster dredging and some of the finest boats in the fleet have started life as pilchard drivers under lug and mizzen. The best known of these are the 'George Glasson' and 'Stella'.

GEORGE GLASSON

Setting the spinnaker on board 'George Glasson'.

The ‘George Glasson’ was built at Portleven in 1898 as a ‘Porthleven half boat’. These were luggers decked to foremast and would have worked as a pilchard driver and long liner. She was bought by Matthew Laity of Flushing in 1903 and used as a lugger until 1923 for trawling, lining and netting. In 1923 she was converted to gaff and probably then started dredging. After Matthew she passed on to his son R.J. Laity and in turn his grandson Paul who dredged her until his death in 1980. She is a big boat at 30′8″ and one of the best known of the local boats. Her present owner Pat Crockford has all but completely rebuilt her and she is now in beautiful condition.

Jimmy Williams new boat ‘Stella’ is launched as a lugger in 1909, but most people remember her better as a gaff cutter owned by Ray Frost.

STELLA

By comparison 'Stella'is only 25′, but one of the fastest boats ever to sail in Falmouth. She was built by Jackie Bawden (Porthleven) in 1909-10 for a cost of about £60 using pitchpine left over from a yacht built for the Earl of Delaware. She was launched as a lugger for Jimmy Williams for crabbing and later for drifting and longlining, and was named after his daughter. She started dredging in 1941 when Charles Trebilcock bought her. He sold her to her present owner Ray Frost of Feock in 1956 who dredged her until he sold her in 1989. He was the first dredgerman to try a terylene mainsail when he put one on 'Stella'. She has a finer transom than most of the Porthleven built boats which tend to have full transoms in some cases extending below the waterline before coming in. Most were built like this for carrying a big mizzen which required a boat to be fuller in the stern. By contrast the Mevagissey boats, though luggers, were much finer aft as they had no mizzen.

BOY WILLIE

Another Porthleven boat built with a fairly fine transom was the 33′ 'Boy Willie' also owned by Ray Frost. Again she was built for pilchards about 1883 and was unusual in that she dredged as a ketch, the reason being that at 33′ a cutter rig would have been too heavy to handle with ease for dredging. She spent some of her life crabbing in the Scillies in about 1918 and in 1961 was bought from divers at Newlyn by Ray Frost

'Gomer' Hitchman hauling a dredge on 'Boy Willie'.

and converted for oyster dredging which she did under several owners until 1973. Ray sold her in 1967 to John Prior, and she returned to dredging in 1969 when Ted Hitchman and Nigel Hunt bought her. They worked her until 1974 when John Walls bought her to use as a motor boat fishing for mackerel and longlining. He sold her in 1976 and since then she has been a yacht. Another unusual feature as a working boat is that she was flush decked. At some stage during the war her bow was strengthened and a gun was placed on it.

MYRTLE

Similar to 'Stella' was the 'Myrtle' she was possibly her sister ship. She was built in 1899 as the 'Eva' for crabbing. She, like 'Stella', had a very full round bilge. When she first came to Falmouth owned by Jacob 'Ferryman' Laity she worked on oyster dredging with the dipping lug instead of gaff, and could apparently work closer to the wind than the gaffers, though this benefit was outweighed by the disadvantages. After Jacob Laity she was sold to 'Father Dicky' Vinnicombe, then Arther Randle and finally finished dredging in 1936. She became a yacht after this and went to a port on the South Coast where she was burnt on the beach. She had a reputation as a fast boat, which is very likely if she was 'Stella's' sister ship.

LILIAN & DOROTHY LOUISE

Two boats built by Oliver at Porthleven were 'Dorothy Louise' 1897 and 'Mayflower' 1903 as probably were 'Lilian' 1902 and 'Lucky Boy' 1890.

David Laity and son Kevin heading for the fishery in 'Momento', ex 'Dorothy Louise'.

'Lilian' on her moorings in Falmouth, ahead of her lies 'Seagull'.

'Dorothy Louise' was built as the 'Momento' and came to Falmouth in 1926 when her name was changed to 'Dorothy Louise'. Just to confuse the issue, prior to her coming to Falmouth in 1940 'Lilian' was called 'Dorothy Louise'! Both boats were unusual as they had a raised foredeck to increase the size of the fo'c'sle.

LUCKY BOY

'Pathfinder' when she was 'Lucky Boy' and owned by the Osbornes, who are sailing her in this photo.

So has also the 'Lucky Boy', and she too had had several changes of name, beginning life at Porthleven as the 'Pathfinder'. She came to Falmouth under the ownership of Edward Hendy still as the 'Pathfinder' and in 1937 her new owner Alec Pengelly also kept her name 'Pathfinder'. Whilst owned by the Osbornes and Martins successively she was known as 'Lucky Boy' but when sold to Mission to Seamen her name was changed to 'Flying Angel'. Until recently she has been mackerel fishing under motor as 'Pathfinder' again.

MAYFLOWER

Probably best known of these four boats is the 'Mayflower' bought from St. Ives in 1937 by a well known waterman Jim Morrison. He worked her until the mid 1970's and still owns her though he no longer works her.

Reefed down for poor weather, Jimmy Morrisson leaves his moorings in Flushing in 'Mayflower'. It will probably take him 45 minutes to sail to the fishery before he can start his days work.

The only five other boats that came to the fishery that are definitely known to have been built at Porthleven were the 'Pet',' Melita', 'Gypsy', 'Pelican' and 'Fearnot'.

GYPSY

The 'Gypsy' was built there in 1900 as 'Chaidee'. She was first known of in Falmouth about 1946 when her owner Bert Moore did her up as a yacht and changed her to 'Gypsy'. He used her for charter cruises and

'Gypsy' as a charter yacht before Clinton Powell bought her for dredging.

she went as far afield as Dartmouth and Exeter on her cruises. Next she was sold to a Mr. Harley in part exchange for a motor cruiser called 'Whaup', and he lived on her until she was bought by Clinton Powell and started oyster dredging in 1962.

MELITA

The 'Melita' was first registered in Falmouth in 1903 by three brothers, William, James and Bennet Dunstone and stayed with the family until sold to R.Eddy of Flushing in 1913. He sold her in 1918 to Herbert Monk of Flushing who used her until 1942 and had her topsides raised by one plank in order to carry oysters to the Duchy Oyster Co. at Helford. In 1949 she was bought by Stan Ferris of Mylor who dredged her until she was sold again in 1974. She is now at Martin Heards boatyard at Tregatreath where she is being rebuilt.

'Melita'.

PET

The 'Pet' first appears in Falmouth register in 1895 under J. Dunstone. No more is known until 'Arfy' Treneer bought her derelict from Helford and converted her to a gaff cutter and took her oyster dredging. She is typical of many of the larger boats in that he would take her mast out in the summer and use her for other forms of motorised fishing.

'Pet', a big boat for 'Arfy' Treneer to work single handed.

PELICAN

'Pelican' was built in 1912 as the 'Daily Mail', a lugger. She was 25′6″ and owned by H. Tregaskis of Flushing and stopped dredging in 1927. 28′ 'Fearnot' was older, though the actual date is not known and finished dredging in 1947 when she was sold to be converted to a yacht. She too was probably a lugger originally, and this boat sadly ended her days by breaking her legs and died where she fell.

ELLEN

Another boat that is supposed to have been built at Porthleven was the 'Ellen'. She was found on Turnpike beach about 1900 by Cyril Bennet's father where she lay on her side having been run down in the harbour. He bought her for 30/- intending only to use her keel but found the bottom to be sound and so rebuilt her. She was fastened with trenails as it was claimed she was built in 1817 and the oldest working

'Ellen' with a short mast only is rigged for trawling. Note the shipping laid up on Mylor Bank.

trawler in the South West. Cyril Bennet dredged her until selling her about 1955 to 'Toby' West. When she got too bad for trawling, she was taken by the RNLI with the intention of setting fire to her at sea for a film they were making, but in the end it was called off. Her keel was sold to go on 'Swallow'.

Probably many other Porthleven boats did in fact work on the fishery, but we cannot say for sure.

Boy Willie

This 'Boy Willie' which was also built in Porthleven is a cutter unlike the other 'Boy Willie' which is a ketch. George Vinnicombe used her as a motor boat for trawling only until recently, but his son Tim decided to fit her out as a dredging boat. The deep gunwales seen in photo 2 have been removed to make it easier to get the dredges on board.

Mevagissey Boats

Eight boats are known for certain to have been built at Mevagissey and to have joined the Truro fishery. They are 'Shamrock', 'Magdalena', 'Winkle Diver', 'Zona', 'Boy Willie', 'Julie', 'Kathleen' and 'Water Hen'.

SHAMROCK

'Shamrock' was built there in 1908. Little is known of her until 1940 except that she was fishing in the Scillies as a lugger at some time. She was brought to Feock in 1940 by William 'Shaver' Ferris, was sold to Coombe in 1952 to S.J. Ferris and she stayed there dredging for Johnny Old who bought her in 1967. After his death she was sold and is now a yacht at Falmouth.

'Shamrock' is almost certainly the only boat on the fishery to have carried a suit of turquoise sails, unfortunately this black and white photo doesn't do her colour scheme justice.

MAGDALENA

'Magdalena' was built around 1890 as 'Helda' and was purchased by J. Soloman of Flushing in 1903. Her next owner in 1924 was an oyster bailiff on the Falmouth Fishery who lived in Flushing – William J. Gilbert but was unfortunately drowned from her whilst trying to save a punt that had broken adrift. After this tragedy in 1933 she went for a yacht and her name was changed to 'Lena'. Her name was 'Magdalena' when she was bought by Clive Warrillow in 1969 who took her oyster

'Magdalena', the dredging boat 'Arfy' bought to replace 'Pet'.

dredging again and sold her to her present owner William 'Arfy' Treneer in 1970, and he is still dredging with her. 'Arfy' had an accident with a saw about 1985 which lost him the end of a few fingers. This made life difficult for 'Arfy' as he could no longer grip the dredge warps properly to haul the dredges up, so he adapted a punt wink for use on 'Magdalena' and mounted it in the centre of the cockpit. It was slower than hand-hauling, but at least he could still dredge within the bye-laws as the winch was hand operated.

ZONA

The 30′ 'Zona' was built at Mevagissey for trawling and longlining and in 1947 was owned by F. Trevarton. He sold her to Arthur Randle in 1959 and in 1965 was purchased by William Tregaskis and Raymond Instance who dredged her in the winter and fished with her in the summer without the mast. Now she is owned by Frank Vinnicombe and dredged by his son, Chris.

'Zona'

WINKLE DIVER

'Winkle Diver'is not a typical working boat as she is smaller being a Mevagissey Tosher. She is 21′ ketch rigged and built in 1934 as 'Tesser'. She was converted for oyster dredging in 1945 by Cyril Green and then sold to Geoffrey Hitchens also of St. Mawes in 1961. He kept her dredging until the early 1970's but she has not worked since then.

'Winklediver'

JULIE

'Julie' was another Mevagissey tosher, and worked on the fishery for a short time.

'Julie' sailing across the oyster banks under working rig.

The 'Water Hen' was built by Pill and was probably very like 'Magdalena' but with a squarer forefoot. She may have been built for the Johns family of Flushing, she was later owned by James Rees and then Mathew Laity of Flushing before she was sold in 1928 to T. Hoskins of Helford. She was later sold 'up country'.

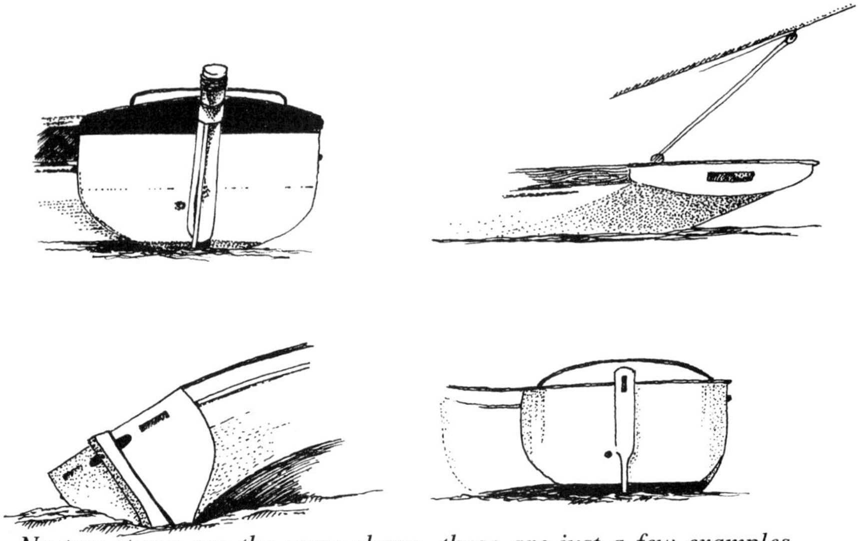

No two sterns are the same shape, these are just a few examples.

St Ives Boats

The St. Ives-Newquay area provided a few boats for the fleet. Newquay being a holiday resort had a fleet of sailing pleasure boats, which also did a bit of fishing. As they were of the hull shape required for oyster dredging three of these vessels were purchased for that purpose, namely 'White Heather', 'Honour' and 'Phantom'. As pleasure boats they were usually licensed for about 12 passengers, had a loose footed mainsail and a long tiller which could be operated from well forward, almost to the mast. The passengers used to stand behind him and the skipper would explain how it was sailed. They carried fidded topmasts and lead keels. When the mainsail was to be lowered, they would lower the gaff and lash the mainsail to the mast. They were about 24′ long.

White Heather

One such boat 'White Heather', was built there in 1905. She was bought for oyster dredging by William Hitchens of Feock in 1919 who put an engine in her a year later. In 1931 Samuel Old of Kea bought her and kept her until 1942 when she went to William Dunstan of Carnon Downs who swopped her for 'Fearnot'. He kept her until 1945, sold her to William Searle at Feock who sold her to be used as a yacht by A. Wilcox until 1953 when she started dredging again under her new owner Frank Hitchens. In 1961 she came to Falmouth bought by Pat Crockford who deepened her and did her up. After him she was sold to Roger Eggins who still continued to dredge her until she was sold to Derek Eury who took her to Guernsey and renamed her 'Petit Pomme'.

She has since been resold and in 1976 again returned to Falmouth for a brief visit with her new owners, P. Boyce of Dorset, who have changed her name to 'Little Apple'.

No wind, no engine. Pat Crockford does the only thing left in 'White Heather'.

HONOUR

Another boat to come from the same area was 'Honour', built by Painter (St. Ives). She was bought from St.Ives by Pat Bagnall who worked her for a while before selling her to Fowey. Frank Vinnicombe brought her back to Falmouth and in 1972 sold her to her present owner Mr. Gunn at Coombe.

'Honour'

ENDEAVOUR

The best boat to come from St. Ives and one of the best of the Falmouth Working Boats fleet was 'Endeavour'. Built in St. Ives in 1911 as a lugger, she was purchased for dredging from Mousehole in 1925 by John Merrifield, and sold to his brother Henry Merrifield in 1968 when he sold his own boat 'Winnie'. Even in his 80's he was still dredging in her single handed. The fact that she sails so fast is even more of a credit to her builder when you consider that she had no external ballast keel.

John and Henry Merrifield on board 'Endeavour'.

RIPPLING WAVE

'Rippling Wave', ex 'Shagreen', was built in St. Ives in 1902 and was bought by Josiah Hitchens of Feock in 1918. She stayed with various members of the Hitchens family until 1930 when Cyril Pascoe bought her. He sold her again in 1935 to Gordon Martin who kept her for one year and then sold her to be turned into a yacht. In 1946 she was bought from Lt. Col. Cornish of St. Mawes by Bart Moore as a wreck, having been run down and sunk by the ferry St. Germans on her mooring. She was lifted and offered as she lay at Polvarth so Bart Moore purchased her off Lloyds agents and repaired her and sold her to A. Harvey of Falmouth in 1947. She finally went away up country and has not been seen since.

The 'Rippling Wave', another boat saved by Bart Moore.

Falmouth Boats

Though many men from Falmouth dredged they did not have many of their boats built in their own port. The emphasis in Falmouth was on quay punts but nevertheless several boats were built at Falmouth, but for some reason none survived in the Fishery later than 1929 and so little is known about them.

The 'Skandinvia' built in 1889 and owned by Edward Johns of Flushing was laid up in 1921 as was Cyril Bennets 'Rover' built 1869. Both these boats had mizzens. Another dandy rigged boat was 'PC' built in 1886 and was no longer fishing by 1927. 'Wild Cat' was lost to the fishery even earlier in 1904, 'Lorna Doone' in 1927, 'Heartsease' 1917 and 'Edith Annie' in 1928 with finally 'Blanche' in 1929.

It seems a little ironic that as the boats are known as Falmouth

Working Boats so few of them were built at Falmouth and none have survived in Falmouth itself.

Five other boats were built in the Falmouth estuary; two at Penryn, one at St. Mawes, one up Cowlands Creek and one at St. Just. The 'Alf Smithers' and 'Three Brothers' were built at Penryn, 'William Green' at St. Mawes, 'Deacher' at Cowlands and 'Queen of the May' at St. Just.

THREE BROTHERS

Built 40 years before the 'Alf Smithers' at Penryn was the 23′ 'Three Brothers' in 1928. What her original name was is not certain for whilst owned by Gordon Oliver of Portscatho as a yacht she went under the name of 'Petrel'. Some time after this she was renamed 'Lady Beatrice'. Bobby and Jan Truscott bought her and in 1966 used her for dredging

'Three Brothers', ex 'Lady Beatrice', ex 'Petrel'.

and in 1969 she was dredging again for Dave Timmins of Penryn. Those were probably the only two years she dredged and she was then sold to Vernon Parker and eventually broke up alongside the quay in Penzance harbour in September 1973.

WILLIAM GREEN

Another old boat no longer in the port is the 'William Green' built at St. Mawes in 1864 presumably for William Green who is her first known owner and had her rigged as a lugger for drifting and lining in 1883. She started oyster dredging in 1906 when Joseph Green had her dandy

'William Green'

rigged for a time, but later in the same year he sold her to Joseph Johns at Flushing who cutter rigged her. He worked her until 1933 when she was sold as a yacht to G. Muirhead at Mylor and remained there for two years. It is not known what happened to her after that except that she was owned for a time by Irving Green at St Mawes again.

DEACHER

Less is known of 'Deacher' which was built at Cowlands and spent some time registered as a Plymouth fishing boat (PH 141). In 1920 she was owned by J & S Martin at Malpas who sold her to William Passmore at Philleigh in 1921. He worked her until 1952 and then sold her for a yacht.

QUEEN OF THE MAY

The Pasco yard at St. Just which is still building punts, built 'Queen of the May' in 1906 for Flushing man Charley Monk. She was designed by George Nicholson Snr. on the lines of the Itchen Ferry boats. Charley Monk was the captain of a coasting schooner and 'Queen of the May' was a gift to him on his retirement from the schooners millionaire owner, Mr. Milburn. He used her to tow barges of oysters to Helford

River. When the Vinnicombes 'Boy Willie' sank 'Queen of the May' and 'Ellen' lifted her, one on each side. She was launched by Charley Monk's daughter Florence May, which may provide a clue to the name 'Queen of the May'. She stayed at Flushing with the Monk family until 1935. In that year she was owned by Francis Vinnicombe at Falmouth until he sold her in 1958 to Gladys Pike who sold her in her turn to R. Martin. She left Falmouth about 1966 when R. Martin sold her to an

'Queen of the May'

owner in Percuil who used her as a houseboat and changed her name to 'Andromeda', keeping her until 1971 when she returned to Falmouth still as a yacht and was owned by Joe Smith. He did quite a bit of work on her before selling her a couple of years later to Fowey where she still is. She is said to be very stiff sailing upright, but throwing an unusually large bow wave due to her big shoulders.

Four other boats to join the fishery came from the nearby ports of Helford and Porthallow, Polperro and Portscatho. They were 'Happy', 'Maud', 'Elizabeth Mary' and 'Seabird' respectively.

HAPPY

The 'Happy' was built originally for the Vinnicombes as the 'Happy Family' at Helford in 1896. She remained with them until 1906 when she moved to Feock as a 21st birthday present for Chas Ferris and at this time was fully decked and probably had a loose footed mains'l as she had a bonnet two feet deep laced on the foot. (A bonnet was a piece of sail laced to the foot of a sail to increase the area). It may have been whilst he owned her (up until 1945) that William Ferris (son of the famous builder) rebuilt her, and as a result she was left with two prominent bulges on either side of her bow just forward of the mast and as Norman Ferris said to me "It looks like she has got toothache!" She was sold to another Ferris after Chas Ferris for in 1945 Ben Ferris of Carnon Downs

The 'Happy' shows the result of her 'dental treatment'.

bought her and worked her up until 1970 when she was sold to Falmouth owner Bob Pointer, who had her worked for one season before he sold her and she finally left Falmouth to go to Portland about 1973. At some time in her life she was owned by two other people and although it is not certain it most probably was Chas Ferris who sold to either Tom Symons or William Cock, and one of these two sold her to Ben Ferris. William Cock lived on her for some time as a houseboat at Malpas, apparently with a ginger Irish Terrier which usually sat on his shoulder. At one stage of her career she was owned by one of the Vinnicombes who came near to losing her when she was struck by a severe squall on the Mylor Bank, filled up and sank. His son, 'Bridly' scrambled to the safety of the masthead (as luckily she was in fairly shallow water), shouting, "I shall be killed by my father"! Fortunately all ended well, the boat was refloated and worked for many years after the incident.

Gorran Haven Boats

MILDRED

R.D. Pill at Gorran Haven built boats mainly for the Mevagissey area but four managed to find their way to the Truro oyster fishery. Two of these are still in the port, though neither have dredged for the last few years. 'Mildred' was built by him in 1913 for William Johns of Flushing. She is one of the larger boats as she was built for trawling and lining as well as dredging, and for that reason had an engine from new. She is 27′6″ long with 8′ beam and more than the average draught, being 5′6″ deep. William Johns fished in her until 1934 when the celebrated author Howard Spring bought her for use as a yacht. At that time she had just had a new bermudan mast put in her and a new Thorneycroft engine, though why William Johns should have put a bermudan rig on her is something of a mystery as Howard Spring's son David Spring says she was a lot slower with that rig than the other gaffers. It was Howard

'Mildred' laid up for the winter up Mylor Creek.

Spring who gave her the fo'c'sel that she now has with bunks and toilet as well as the two portholes which few working boats have, as it enabled him to sleep aboard in greater comfort when cruising. He owned her until 1962 and then sold her to Alec Bailey who sadly died at the helm whilst cruising off the Yealm and the boat finished up on the rocks. She was sailed back to Falmouth by Bart and John Moore and sold to Russel Myners who kept her for only six months before selling her to Peter

Davies at Mylor. He did a little dredging in her but mainly used her for cruising and then sold her to Mr. Roebuck as a yacht again. She had a narrow escape in 1968 during the gale that sank the 'Six Brothers', as she too broke her moorings and went ashore. She was saved by Clinton Powell, Mike Parsons, Paul Laity and David Annear, who managed to get on board, haul a jib and sail her off stern first. Her current owner David Stonehouse has had her since 1975 and uses her for racing only.

ORLANDO

The other boat built by R D Pill and still in Falmouth is the 'Orlando' which he built in 1911. She came to Feock in 1944 when Bernard Williams bought her for dredging and stayed there after John Mills bought her from him in 1959 and until he sold her to Falmouth in 1960 where she still is.

BOY JIM

The 'Boy Jim' was built by R D Pills', father a few years before the 'Mildred' and 'Orlando' some time before 1894 as the lugger 'Ellen' for drift netting and long lining. She was bought in 1894 by J. Bight from two brothers at Percuil who fell out and so he got her for only £5. She remained his until she was broken up in Penryn in 1931. She must have been quite fast as she beat 'Queen of the May' and 'Morning Star' at Flushing Regatta, though she only raced once.

FAIRY

The only other boat imported from Gorran Haven and also built by Pill was the 25′ 'Fairy' which Tom Symons owned between 1905 and 1917 when she was broken up.

Boats from Other Ports

MAUD

'Maud', ex 'Lilian', belongs to the list of Falmouth working boats by virtue of dredging in 1965 when Rodger Eggins fished her, and again in 1971 and 1972 when Peter Divers fished in her. She was a 21ft. clinker built Porthallow crabber with a dipping lug built originally

'Maud' on her moorings off Little Falmouth. One of the few clinker built dredging boats.

with a centreplate, and it wasn't until Rodger Eggins bought her that she had a keep put on her. Before Rodger Eggins owned her she belonged to a John Smith and after Rodger Eggins to a Falmouth postman, and then finally to owner Peter Divers, her last known owner who sold her to 'Murre'.

SEABIRD

In 1890 the 21ft. 'Seabird' was built at Gerrans for Harold Nichols of Portscatho for lining and oyster dredging. She was a clinker built boat and he kept her until 1910 when she came to Falmouth owned by N. Williams who in 1914 sold her to F. Harmet who owned her until 1922. Then she was sold to William Ellis at Ponsmains who worked her as did his son from 1934 until he sold her to Ernest 'Brother' Searle in 1937 who took her to Christchurch eventually, but used her at Falmouth for a time taking out trippers.

ELIZABETH MARY

The 'Elizabeth Mary' is one of the few surviving Polperro gaffers, and she was dredged between 1965 and 1970 by her owner Mike Johnson. Since then she has been sold to Dartmouth where she still is. From 1920 to 1965 she was used in a sailing school run by Lt. Col. Patrick O'Hara Phibbs O.B.E.

'Elizabeth Mary' at Fowey.

ZIGUENER

Of the remaining boats whose place of origin is known, perhaps the one with the most interesting history is the 'Ziguener'. She was not, as is thought by many, built at Restronguet, but only rebuilt there about 1840. 'Ziguener' the name of a German gipsy was probably a German yacht originally and was bought and converted to a dredging boat by William Ferris about 1840. As a yacht she had too deep a keel for dredging so William Ferris cut her off at the waterline and replaced her clinker bottom with a shallower carvel one whilst she was balanced on barrels on the beach. Her first entry in the fishing register was in 1886 when she was owned by George Vinnicombe of Falmouth for long lining and her next appearance was in 1906 when owned by George 'Matey' Vinnicombe of Mylor and used for oyster dredging. He owned her until 1918 when she went to Restronguet and her next owner John Ferris. She was eventually laid up in 1951 and fell to pieces in Restronguet Pond, the local name for the little dock near the Pandora Inn. She is seen in photo page 43.

LILY

Another boat to travel a long way in order to dredge Falmouth oysters was 'Lily' of Kings Lynn. She was built in Norfolk in 1906 for oyster dredging and trawling off Kings Lynn. In the mid 30's she went to the Norfolk Broads and was converted to a yacht. During the war she was stationed on Oulton Broad as a mooring lighter until about 1954 when D. Sayer bought her. He used her for cruising after he had re-rigged her, and during the next ten years he kept her and cruised as far away as Holland. Her next owner named Bellamy gutted her, intending to do her up but sold her in this condition to a South Crofty

Though designed to work off Norfolk, the pretty 'Lily' adapts easily to the Fal.

miner and eventually she was sold to Bart Moore in 1970. Though she was not typical of Falmouth working boats Bart decided that she would make a good dredger, regardless of the fact that she was 35′ and had a counter stern. In spite of setbacks, the collapse of such a new counter Bart was putting on her whilst it was propped up with cramps on the beach, (he just picked up all the bits and started again) his faith in her proved to be justified as she dredged very successfully for his son John for three years. He worked her until 1973 when he bought 'Katherina' from Derek Veevers.

MOYA

Probably the most extreme hull shape that has been seen on the fishery was that of 'Moya', a Morecambe Bay prawner built by Arneside about 1905. She was typical of all Morecambe Bay prawners, being very beamy, shallow with a rounded counter stern. She is 34′ long, 11′ beam and 3′6″ draught. Her construction is pitchpine on grown oak frames. However her owner David Cain found that she was not 100% satisfactory for the job as being shallow and beamy she did not grip the water very well whilst drifting and tended to skim across the water and

Not a great success as a dredging boat due to her very shallow draught, 'Moya' did not work for long.

drift too fast. However he used her for three seasons 1967-70 before she was sold as a yacht.

SEAGULL

Though the 'Leila' was designed on an Itchen Ferry Punt, the only authentic Itchin Ferry Punt to work in the fleet was 'Seagull'. She is

21′ and has a large beam of 8′6″. She was bought in 1969 from a doctor at Redruth by Tony and Robin Arthur at Flushing who converted her for oyster dredging though sold her before they had a chance to use her themselves, and it was her next owner Peter Lockrie who first put dredges aboard. Whilst he owned her she was called 'Dylan Thomas' but he only dredged her for a short time and the next year she was sold to her present owner Graham Pearce who has dredged her for two or three seasons before replacing her with his 28′ GRP boat, 'Cousin Jack'. Finally two boats of which the place of origin is not certain are the 'Morning Star' and the 'Ripple'.

The Itchen Ferry Punt 'Seagull' shows her forefoot as she rides a sea.

MORNING STAR

The 'Morning Star' was, until recently, the oldest boat still surviving on the fishery dating back to at least 1830. She was built for a Portloe fisherman possibly A. Trudgeon who owned her all his life and died at 93 in 1890. He used her as a seine boat and was probably lugger rigged. After he died she came to Mylor and lay derelict there for some time possibly owned by one of the Ferris's until 1906 when William 'Mubber' Harris bought her. He had her lengthened at Coombe by a boatbuilder called Wellington who was also landlord of the Punchbowl Inn. He lengthened her by 6 or 8′ by the bow to 33′6″ and then worked her until 1955 when she was bought by Frank Cock and Harry Barnes. When Harry retired Frank continued to work 'Morning Star' with his son John, until she was run down on her moorings in Restronguet by an unknown boat one night in 1980. She was damaged beyond repair and is now an exhibit at the 'Lands End Leisure Centre'.

Frank Cock does a few running repairs on the 'Morning Star' on the beach outside the Pandora Inn.

RIPPLE

'Ripple' is alleged to have been built at Mylor but as she has a counter stern it seems rather unlikely, possibly she may have been built as a yacht or been a fishing vessel from the east coast of England where counter sterns were far more common. She was supposed to have been built about 1886 but in the Truro boat register there are two entries one in 1875 and one in 1869, both of which appear to be her. In the first she was owned by William Gay of Pill and in the second by Chas. Ferris also at Pill who bought her in 1874. The next record, again in the Truro Fishery register is in 1906 under the ownership of Thomas 'Bummer' Ferris of Mylor who in 1917 moved to Falmouth and the boat was re-registered at Falmouth under his ownership and stayed so until she was struck off 'no longer fishing'.

As well as the boats built by P. Ferris for the fishery, three other boats eventually found their way to the Fal from Looe. These were 'Sunbeam', 'Twilight' and 'Vivian'.

SUNBEAM

The 25′ 'Sunbeam' was built at Looe in 1912 as 'Phyllis A' and rigged as a cutter for longlining or drifting. She was first registered at Truro under the ownership of James Lewarne who kept her until she was sold out of the port in 1951. However she was soon back and in 1953 she was re-registered at Truro under her new owner J. Rowe of Mylor whose family still own her, though she now lies derelict in Mylor Creek opposite Tregatreath.

'Sunbeam' ex 'Phyllis A'.

TWILIGHT

The larger 36′ 'Twilight' was built at Looe in 1903 as a lugger for longlining and drifting. Prior to the second world war she was based at Newlyn owned by the Stevenson family. During the war she was laid up and in 1947 was condemned, but was saved by virtue of being too rotten to take out of the harbour and sink! In 1959 she was bought by Barry West (Falmouth) and he dredged her between 1966 and 1970, ketch rigged and with wheel steering. That was the last time she dredged and she was sold to Paul Wherry. He sold her to Stan Judd at Mevagissey, she came back to Truro when he sold her to Russel Myners, then to J. Fletcher at Cadgwith and she finally left the port for Scotland where in all probability she still is.

'Twilight' alongside Flushing Quay.

VIVIAN

Yet another boat to undergo several changes of name was 'Vivian'. She was built at Looe under the name 'Bessie' and came to Falmouth in 1919. In 1944 B. Moore bought her as the 'Winifred' from a Mr. Garner. She had fallen on her side and was in need of some attention and having repaired her he sold her to G. Collins (Mylor) in 1945. Her next owners were Messrs. Knight & Grice, who in their turn sold her to Jennings at Flushing; then to Terry Heard at Mylor in 1955. Once again Bart Moore bought her and did her up selling her to Douglas Martin (Falmouth) who renamed her 'Vivian' after his brother. (See photo page 15.)

MAIDEN NELLIE

From Mousehole came three boats, 'Maiden Nellie', 'Boy Phil' and 'Boy Peter'. The 'Maiden Nellie' had the reputation of being one of the fastest boats to sail at Falmouth. Built as 'Boy Philip' in 1903 at Mousehole as a crabber, she was bought by William 'Scottie' Anderson and brought to Falmouth in 1924 and renamed 'Maiden Nellie' after his daughter, he kept her oyster dredging until 1944. In 1945 William Dunstan bought her, when he bought her she was under water and in need of a lot of repair as a result of going aground on the wreck of the 'Tuscalusa' off Greatwood Quay where she remained for two weeks

'Maiden Nellie'

before she was taken off. He put in new decks, beams, planks, floors and replaced the original steam frames with grown ones using sawn oak from Carclew. He kept her dredging until 1952 when she went to Fowey. After this in about 1953 an attempt was made to sail her across the North Atlantic by the old Viking route. She was Bermudan rigged during the attempt and left from Stromness but as she started to leak badly, the three men on board (David Reagen a journalist, Leslie Shephard a film writer and John Fielda Cambridge graduate in engineering) decided wisely to turn back.

BOY PHIL

'Boy Phil' was built at Mousehole as 'Jubilee' in 1897 and was lug rigged and probably a crabber. In 1918 she was bought by Francis Symons of Feock who renamed her 'Boy Phil' after his son. Next she passed to Francis Lewis Gunn at Kea in 1932 and in 1935 on again to Gordon Martin. He appears to have kept her for only 3 months before selling her to H. Laity at Falmouth. He sold her to St. Just until Frank Vinnicombe bought her and worked her from Mylor after this and in 1975 Lesley Angell bought her and it was he who sold her to the present owner John Walls of Flushing.

Roy and Rex Collins dredging on Frank Vinnicombe's 'Boy Phil' before they built 'Rhoda Mary'.

BOY PETER

Not much can be said about 'Boy Peter' which can only be classed a 'Falmouth Working Boat' by virtue of having been dredging. It was certainly not typical, being a 26 foot clinker built crabber of very little draught as can be seen in photo 20. Arthur 'Shoeboot' Rowe bought her from Mousehole and in 1949 Bart Moore put on a wooden keel to deepen her for sailing, but it was not a success and after one season he gave up with her.

ROYAL OAK

Yet another boat from Mousehole, was the 'Royal Oak', 24′, built by Paynter in 1902 and not to be confused with the 'Royal Oak' Hitchins built, now 'Victory'. Like many of the boats to have come from this area she has steam frames. First record of her in the Falmouth area was at Port Navas when a Mr. Graff owned her from 1925-1956. He sold her

'Royal Oak' (left) and 'Muriel' (right) on the beach at Pill Creek.

to another person from whom a Mr. Sinfield purchased her about 1961 and kept her until 1968. During this time Bart Moore did some fairly extensive repairs to her before she was sold to Terry Major at Pill who used her dredging between 1969 and 1971. He sold her to, and after his death she was dredged by Nigel Molin who sold her to Roger Martin in 1987.

MOLLY

In 1979 the market appeared to be booming for fibre-glass Working Boats and so the local GRP fishing boat specialists 'Cygnus Marine' got Percy Dalton to design a 28′ boat for them to produce. In 1980 Robbie Arthur of Flushing purchased the first of these and fitted her out himself and named her 'Molly'. She proved to be a very able boat, though to date she is the only one of them to have worked on the Fishery. When Robbie emigrated to Australia the boat was sold to Sid Goodwin of Falmouth who put a cabin on her showing what fine cruising boats they can make when their dredging days are over.

Robbie Arthur racing in 'Molly'.

COUSIN JACK

The last boat to have been built of fibre-glass was the 28′ 'Cousin Jack'. When Graham Pearce sold his Itchen Ferry Punt 'Seagull' he replaced her with a boat of quite an unusual construction. Though popular in the USA, 'C Flex' planking never seems to have caught on in Britain, perhaps surprisingly as it offers the big advantage of being able to build a 'one off' in GRP as no mould is required. The boat is laid up in the way a wooden boat would be except that the planks are made not of wood but fibre-glass, and once fixed into the required shape they are then fixed in position with matting and a gel coat. The design of this boat was yet again one of Percy Dalton's and has proved to be a fast boat during her racing career, though as yet she has not done any oyster dredging.

If there were a 'menagerie' section for the strangest boats to have dredged, surely the following boats would be certain of inclusion. Registered in 1869 were the 60′ 'Pioneer' which was spritsail rigged, the 40′ cutter 'Anna', the 45′ cutter 'Rob Roy', the 50′ lugger 'Two Brothers', and the dandy (yawl) rigged 45′ 'Caroline'. In 1870 the 43′ ketch 'Ailsie' was registered for oysters and in 1903 the 38′ 'Silent' (an auxiliary motor yawl built at Mevagissey in 1898) was also registered for dredging.

However, closer inspection of the Customs and Excise records reveal that 'Rob Roy' was listed as an oyster carrier, 'Ailsie' was owned by an oyster merchant, John Mead, and 'Silent' was owned by the Vinnicombe family who were oyster merchants at that time. In all probability the purpose of these boats was the carriage not the catching of oysters, as they were so obviously unsuited to dredging on the fishery.

The 'Cousin Jack' built by Graham Pearce to replace the much smaller 'Seagull'.

CHAPTER 4

Racing

Fishermen have always been and still are competitive by nature, so it comes as no surprise that Working Boat men have developed their racing to the extent that they have done. During the closed season for oyster dredging, each village on the Fal has traditionally held its own regatta, the highlight being the Working Boat race. This has given the skippers many opportunities to exhibit the high degree of skill in boat handling most of them have acquired during the harsh conditions of their winters work. Nearly every week throughout the summer a regatta is held somewhere and nowadays points racing is held by the Flushing Sailing Club on any Saturday when there is no regatta for them to attend.

The regattas are organised by local committees in each of the waterfront villages situated on the creeks of the Fal. A cup is usually put up as the first prize for each class in addition to the money prizes for the first three places. Five pounds for first, three pounds for second and two pounds for third would be typical. In the early 1900's there was one race that was always particularly well attended, and with a prize of thirty shillings for each boat that competed it is hardly surprising that as many

A dredging boat, a Quay Punt and a small yacht prepare for racing – probably they will all race in the same class. The two 18 footers in the backgound will have their own class.

as thirty boats were on the line when the start gun went off. This event was organised by a Mr. Hewitt, who was a resident in the Greenbank Hotel, and it continued up until the beginning of the First World War when finally so many boats were laid up that it was discontinued. In addition to the thirty shillings for each entry there were handsome prizes for the first three boats home. Though many Falmouth Working Boats took part in this race it was not for them exclusively, but was open to all 'working' boats, which at that time meant what it said included any boat that worked for its living such as Falmouth Quay Punts and fishing boats. During the war the local regattas were poorly attended as many of the fishermen were away, but afterward when the war was over and two or three boats started to race again, the Carrick Sailing Club was started in a shed at the mouth of Restronguet Creek, organised by Mr. Norris.

Point and Penpol Regatta, left to right 'Harriet', 'Ziguener' 'Dulcie' (a yacht), 'Shadow' and 'Twinkle'.

An interesting period in the history of Working Boat racing was the 1950's, the time when 'Victory' and 'Maiden Nellie' changed from gaff rig to bermudan rig. Racing at this time was not as readily available as it is today, there were of course the races that were organised for them at the local regattas, but other than that they would have to race at the local sailing clubs against yachts. The situation improved after 1948 when the local sailing clubs got together in an effort to start the week long regatta now known as Falmouth Week, during which there was racing every day except Sunday. This in addition to the usual annual regattas at Mylor, Feock, Flushing, Restronguet, St. Mawes, Porthallow, Portscatho, and Point and Penpol, made for a full calendar of regattas for the racing man, and as most of the men fished in the summer months they could easily arrange their fishing activities to coincide with the racing. For instance,

Possibly 'Shadow' at Point and Penpol Regatta.

Thought to be 'Harriet' (left) and 'Florence' at Point and Penpol Regatta.

it happened on more than one occasion that Henry Merrifield would be fishing off Portscatho in 'Endeavour' on the same day as Portscatho regatta, and given his known fondness for racing, the outcome was a foregone conclusion --- fishing would stop and he would head for the start line. It was undoubtedly a coincidence that he just happened to be fishing in that area on that particular day, but it would be madness not to take full advantage of such a stroke of good fortune! Toby West is another man who has been fortunate enough on many occasions to find himself to be fishing near to the right place at about the right time, though with Toby there is always the lingering suspicion that he might have actually engineered the situation to his own advantage . . . Whatever the truth behind these 'happy coincidences' not many races were missed by him once the racing season got under way, though he has always protested that he was "just a victim of circumstances".

'Victory' in bermudan rig. Not enough wind to keep the spinnaker out of the water.

'Victory' chasing 'Maid Nellie'.

'Victory' with a staysail set as a small spinnaker.

They all raced in 'Class A' in the yacht races which were open to yachts and Working Boats alike, and in 1951 when Ray Frost in 'Stella' joined the Restronguet Sailing Club, he was very quickly joined by the 'Maiden Nellie' and 'Victory', both sporting their new bermudan rigs, and by the end of the year 'Shadow' was there too but in gaff rig. There were other boats racing during this period, though as a rule they only came out for the regattas not the club racing. These included 'Endeavour', 'White Heather', 'Mayflower', 'Harriet', 'George Glasson' and 'Lucky Boy', but undoubtedly the keenest rivalry of this period was between Toby West in 'Victory' and George Dunstan in 'Maiden Nellie' (though she was always known locally as 'Maid Nellie'). George Dunstan bought the 'Maid Nellie' in 1945, racing her until he sold her in 1953. She was bermudan rigged during 1951, 1952 and 1953, and the

The top'sl lies on the sidedeck as 'Stella' beats up river towards Flushing.

Jimmy Morrisson racing 'Mayflower'.

competition between her and the 'Victory' was so keen that most of their races were won or lost by seconds on handicap. Being a smaller boat than the 'Victory' Toby had to give 'Maid Nellie' two minutes a round handicap. Over the years they raced each other 'Victory' comes out marginally on top with a few more wins than 'Maid Nellie'. There was a certain amount of opposition to the boats carrying Bermudan rigs as they outclassed the boats with the traditional gaff rig, and so when the 'Maiden Nellie' was sold, Toby West put the gaff rig back in 'Victory'.

It was about this time when the standards of sails began to improve, the first boat to invest in a terylene sail being the 'Stella' in 1954. Up until now all the sails had been made of cotton or canvas but, terylene offered a real improvement in sails if you intended to use them for racing, and so were very welcome.

Racing in the 50's and 60's was more than just a sport to the skippers and crews, it was a large part of both their social and working life, the two being almost inseparable and yet each an extension of the other. So much did they enjoy it at Point and Penpol that when in 1963 no regatta was put on, the men organized their own. The prizes were denoted by well wishers, put on a table (one for each boat) and they came forward in order of finishing and chose the one they wanted. From their work came the skills and companions with which they were able to enjoy their social life. Racing would always be keen and hard as would the prize giving sessions in the pubs afterwards. Here it was not uncommon for the hard earned prize money to be shared out equally amongst all the boats regardless of their final position, this was particularly common when the boats were in Fowey for their annual weeks racing in the port. If not shared directly it would be shared indirectly by way of filling the winners cup with a lethal mixture of which brandy would normally be a major constituent, though the men sharing it were always prepared to improvise should the need arise! Every boat crew would be there after the racing with its own beer kitty, probably starting with £1 per man only to be topped up with a further £1 each time that nights elected 'treasurer' felt it was getting dangerously low. It was a fair system as all would share equally in the cost of the nights refreshment, the only pitfall being for a slow drinker, as the pace was dictated by the faster drinkers and so once it was time for the next round anyone not ready missed out. Working boat racing trained you in more than racing techniques! The atmosphere was of one large happy family, all the crews would be there as it was unthinkable not to be there if you'd taken part in the race, the talk would be of racing, working boats, fishing and other common interests. You moved freely amongst the crowd in the pub and were able to talk with anyone and everyone as all knew each other and were friends, until at last someone would 'strike sound'. After racing and

The start of a working boat race at St Mawes, left to right, 'Florence', 'Victory', 'White Heather', 'George Glasson', 'Six Brothers' (the original) and 'Stella'.

drinking the working boat mans next love is singing, after all Cornwall is as renowned for it's fine bass voices as is Wales for it's tenors. Many of the men are dedicated choir members who enjoy singing for the sheer pleasure of it. Once the singing started it usually continued until closing time and commanded the attention of all in the pub, working boat crew and others alike who would be drawn to the singing like moths to a light, those who were fortunate enough to possess good voices adding to the sound whilst others who didn't were able to enjoy the practised routines learnt during endless hours of choir practise. Woe betide any unfortunate who made the mistake of talking during a solo, as once the singing starts things are taken rather more seriously!

The atmosphere at a working boat race at these times was that of "a day out for the boys", but they were misleadingly informal as in reality the standard of racing was exceptionally high and very competitive. The apparent unconcern which manifested itself in noisy banter and crates of beer, was only possible as the result of many years of boat handling under sail which bred sufficient confidence not to have to concentrate totally on sailing the boat as that becomes an act of instinct. What may appear to the uninitiated as an act of recklessness does not worry these highly proficient helmsmen, an indication of the confidence they have

Whilst obviously racing, for some reason 'Florence' is only carrying a working jib.

in their own ability to extricate their boat from any situation in which they place it. When you consider that most skippers work their boats dredging and probably spend more hours sailing in one winter than the average yachtsman does in ten years, in conditions never experienced by most yachtsmen, and added to this they race their boats every week each summer, it is hardly surprising that such a high standard is reached. In the pre-war days of the big 'J Class' yachts, it was always such men who skippered and crewed them as paid hands, but with the end of the big yachts and the change over to the more easily handled Bermudan rig instead of gaff, most yachts were now able to be handled by their owners. With such a deep understanding of his boat, acquired by years of dredging under sail, the skipper need only skim the surface of his knowledge of boat handling for racing purposes, though of course he needs to draw fully on his local knowledge of winds, tides, etc., to compliment his other skills.

Because they have such a vast reservoir of experience to draw from in the older dredgermen, the standard achieved by a young skipper is

George Corke and 'Feety' Spear racing 'Evelyn' at St Mawes.

far higher than his counterpart skippering a yacht. The older skippers willingly impart their knowledge to younger skippers (unless it is felt that a particular piece of information is going to cost them too many races), in which case the best that can be expected are a few veiled hints that the recipient will have to work out for himself!

Most of the racing takes place around Falmouth, the only venues outside the Port being Portscatho Regatta, Porthallow Regatta and Fowey Week. During the second and third weeks of August there are two weeks of concentrated racing. Falmouth Week starts the fortnight with Mylor Yacht Club on Monday followed by Restronguet Sailing Club, Falmouth Town Regatta, St. Mawes, Royal Cornwall Yacht Club and Flushing Sailing Club, in that order during the rest of the week.

After Falmouth Week there is a chance to regather the energy expended after the weeks hectic activities, but then on Monday the fleet sets off on a passage race to Fowey. This was probably the highspot of the working boat racing calendar, and so most crews tried to take their annual holidays to coincide with it. Though the trip to Fowey can be a hazardous one should the weather be bad, it would take a little more than 'a bit of wind' to stop the boats getting there, and maybe a bit less to stop them coming back! However, the boats are open boats not designed for heavy weather in open waters and the Dodman Point is not a place to be taken lightly in strong winds.

Once safely at anchor, there is usually time to prepare the sleeping arrangements on the boats. Generally everyone lives on the boats and so a tarpaulin is thrown over the boom and the sleeping bags laid out below. After a hastily prepared meal on board it is time to get changed and go ashore and visit old friends not seen since the previous year. There was a strict rota in which the pubs were visited, though this tradition has died out to a large extent nowadays. A race is scheduled for the working boats every morning of their stay in Fowey, and after the mornings racing is done the real business can continue, which is of course conducted in the pubs! The afternoons are spent watching the local men perform in their spectacular Troy boats. These 18 foot Bermudan rigged keel boats carry a vast spread of canvas for their size, and race every afternoon during Fowey Week, providing not only a source of entertainment but also the chance to win 'a couple of bob' should you select the winner as a draw would normally be organised to add a little spice to the event. Once the race has finished there is just the right amount of time in which to go back on board, grab a bite to eat, and get dressed to go ashore. The working boat spirit is always strong, wherever the men go ashore, but never more so than at 'Troytown' as Fowey is known to the Falmouth men.

The highlight of the week for the working boats is always the

Kenny Corke concentrates on his sails as he leads the race from Falmouth to Fowey as 'Evelyn' nears Fowey.

'Toby' West sails 'Victory' into Fowey with a bit of a sea running.

Harbour Race on Thursday afternoon, when the harbour is cleared of all other boats and the working boats have a race, three times around the harbour, putting on a spectacular sight for the crowds ashore. There are usually large crowds ashore watching as it is part of the Fowey Week Carnival programme, and they usually get their moneys worth too as the racing can be very exciting due to the restricted space in which to manoeuvre, with flukey winds that allow a boat to go from first to last in a matter of moments. At the prize giving on the Town Quay afterwards the winner collects a silver cup and a barrel of beer which is shared between all the boats in the race.

Finally the bags are packed, the boats made ready for sea, and on Friday morning they head for the harbour entrance and the start line for the passage race back to Falmouth. The last the people of Fowey hear of the 'Golden Trumpet' of 'Victory' crew member Steve Whitford is as he plays the last post on the way down through the harbour! Then as the boats head for home the Fowey Harbour tug blows a final farewell and another race is underway.

Another port to be visited in more recent times is Dartmouth, where an Old Gaffers Race was started a few years ago. The first year three Falmouth working boats went up for it, they were 'Evelyn', 'Mildred' and 'Magdalena'. They all sailed up for the race in August 1977 and spent a few days in the port. When the start gun went there were three boats on the start line with spinnackers set – 'Evelyn', 'Mildred' and 'Magdalena' – and that was the order they finished, first, second and third. In fairness to the other competitors it must be said that their boats were designed more for cruising than racing and for them a race is more of a rally than a serious race, but old habits die hard for the Falmouth men who have to race hard as it is the only way they know. The only other trip to Dartmouth after this was by 'Arfy' Trenear in 'Magdalena' a few years later in 1980. Sitting in the Royal Standard pub in Flushing, he decided one night at closing time that he would go to Dartmouth that night and so, fully provisioned with a bar of chocolate and a packet of cigarettes, he set off single-handed for Dartmouth in a 22 foot sailboat with no engine on a journey of 75 miles. Upon arrival he raced the boat single handed and then sailed himself back again the next day. There is a prize given to the boat that has made the most effort to get to the race, which he did not get, as it was given to a boat whose crew had all clubbed together to raise the money for insuring their boat for the race (Arfy had none). Upon hearing this tale from Arfy on his return, one dredgerman felt prompted to say to him 'You'll have to carry 'Magdalena' on your back up the A30 if you want to win that prize!

Another port where the potential of the Falmouth Working Boats has

Racing has always been intense in the working boat races, but this time just a little bit too intense perhaps as 'Winnie' gets her bowsprit entangled in 'Magdalena's' mainsheet, parting the bobstay on 'Winnie'.

been witnessed is the Isle of Wight, where the first race for that most prestigious of all Yachting Trophies – The Roman Bowl for the Round the Island Race – was won by a Falmouth Working boat named, very aptly, 'Merry Conceit'. The Island Sailing Club at Cowes started the race in 1931 and the 22 foot 'Merry Conceit' was the first boat to put her name on the trophy. She was converted to a yacht by then, and owned by Henry Trefusis and Peter Brett, but still traditionally rigged. There must have been some red faces when the little fishing boat with her tan sails took the race against the fancy yachts who were her opposition.

Few boats have returned to Falmouth to challenge the boats on their home ground, but two to have done so were the 'Maid Nellie' and the 'Tern'. 'Maid Nellie' unfortunately did not have the sails of the standard she had in her racing days, for by now she was a cruising yacht. Also she was in need of a scrub and so was not in a position to offer much competition, but the locals were very pleased to see her home again, if only for one race. She had been a very fast boat in her day, some say the fastest of all the fleet.

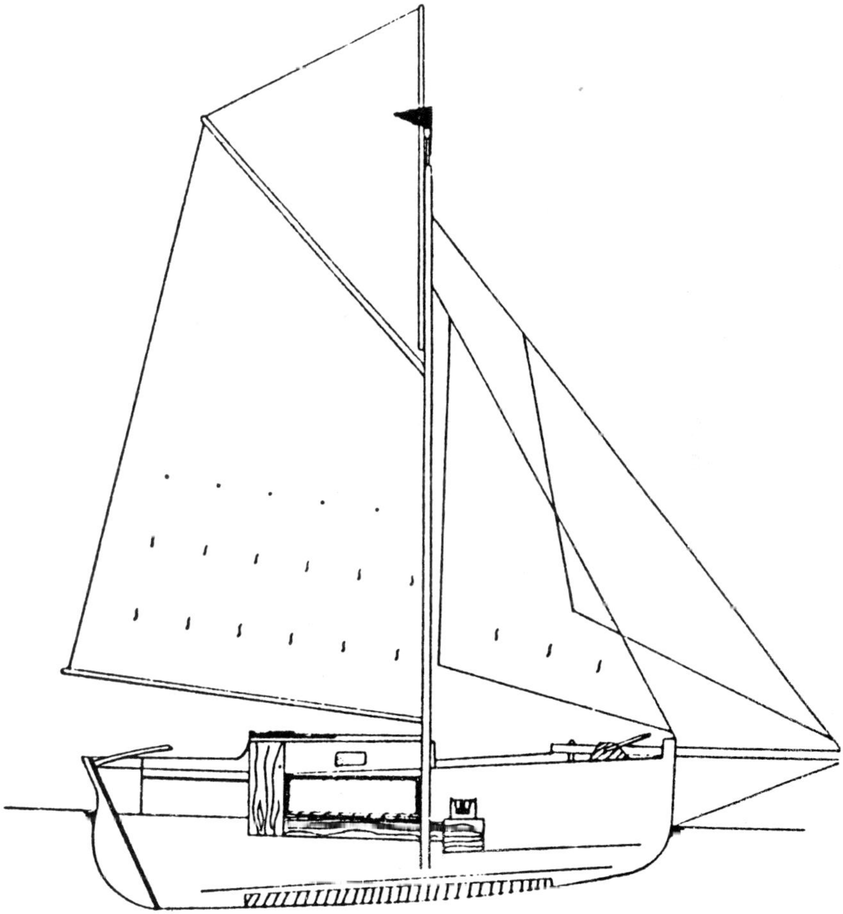

The 'Merry Conceit'

When 'Tern' returned it was slightly different as she came fresh from her victory at the Old Gaffers Race at Cowes, which is a much bigger affair than the Dartmouth Race. She entered two or three races in Falmouth Week 1973 and there was much talk as to what the outcome would be with the 'champion' against the locals. However, she did not have enough speed for the Falmouth boats as the local boats rig had improved a great deal over that period and every boat showed her a clean pair of heels. 'Tern' had a slightly different rig than the locals, as she had a fidded topmast and so had a larger topsail (and a smaller main). So it must be obvious that these boats are somewhere near their maximum as far as their development goes and the only way to improve them appreciably is by the methods that the Association has banned, as they would be to the detriment of the spirit of the class (i.e. alloy masts, lead keels, etc).

'Tern' racing during Mylor Regatta 1971, far right with 'Stella' in foreground.

The mainsail is sometimes the same one that is used for work, but more often an older one is used for dredging, so as to lengthen the life of the racing one. The headsails are far larger and topsails and spinnakers are also set. The spinnakers are huge sometimes as much as 500 cu.ft. and are boomed out on the opposite side to the mainsail on a pole up to about 30ft long. They used to be the old fashioned yacht spinnakers (usually cast offs) but nowadays few of these are left and it is more common to see the new balloon spinnakers made of terylene, and these too are still usually cast offs from modern yachts. To set one, the tack is attached to the outboard end of the spinnaker pole, the head to the masthead or topsail pole and the sheet is led onto the foredeck. In order to prevent the pole from lifting in the puffs one of the crew

Nowadays spinnakers are only seen in photos like this, 'George Glasson' sets two at once!

'Victory' racing in Penryn Regatta

stands on it as far outboard as he dares, and in heavy wind two crew are needed. The topsail is set on a pole of about 15 ft and is supported half way up by the masthead and free standing above it. The heel of the pole is attached to the gaff or the main mast and when attached to the gaff gives the common characteristic in some of the boats of the topsail leaning forward over the masthead. Some boats set jackyards on the foot of the topsail and some merely sheet on to the end of the gaff without one. The first sight of a working boat running downwind is quite breathtaking and is only surpassed by a fleet of them sailing downwind together, for with all sail set all that can be seen is a mass of canvas 50ft. high and spanning 45ft. across with both boom and spinnaker pole out, and a tiny hull somewhere below going about eight or nine knots. Sailing to windward is equally spectacular, as part of the art of sailing a boat to windward is to carry as much sail as possible without capsizing.. Being open boats a sudden gust may put the side right under and fill her up, and with 3 or 4 tons of ballast they don't float! So far only two boats have sunk whilst racing, the first of which was the 'Winnie' in 1971. She was hard on the wind off the Eastern Arm of Falmouth Docks in a strong westerly wind when a gust hit her and she started taking a lot of water. Before the crew had a chance to do anything she started settling by the stern and sank in about 30ft. of water with the masthead sticking out of the water. The crew were all picked up by boats that were nearby, but 'Winnie' had to remain there for several more days before she was raised by local diver, D. Rowe, none the worse for wear, just a little dirtier. Imagine the feelings of a couple who were sailing in from France for the first time just after the sinking and saw what they described as "The most beautiful boat they had ever seen with full sail set under water". After this mishap her owner Joe Keates, decided to put a heavier keel on her with an extra three quarters of a ton and increased her draught by 9 inches to 5ft. 6 inches.

The other boat to sink was the 'Six Brothers' in 1978 during the G. Foulkes Cup. The wind was not exceptionally strong at this time, though very gusty from the N. West. As the boats were in a group heading for St. Just buoy, on the wind, a gust hit all the fleet much harder than the general strength of wind (force 5) that day. What made it so dangerous was that it was a very 'free' puff (i.e. on the weather beam) and so the normal procedure of spilling the wind by luffing would not work as the boat could not turn head to wind fast enough. The same gust hit other boats in the fleet, but it was the 'Six Brothers' that took the worst of it. She sank very fast in under ten seconds in about 100ft. of water with about six hands aboard. Martin Heard, a non-swimmer, had his feet caught in some ropes and was dragged down before he could get himself free and head for the surface. Another, Nigel Mowlin, was trapped

The new 'Six Brothers' just after Dougie Rowe salvaged her, his boat 'Boy Brendan' is just behind her.

under the mainsail and had to dive down several times before he was clear to come up. The third non-swimmer, John Taylor, managed to come straight to the surface and stayed afloat until help was at hand a few minutes later. The race was, of course, abandoned instantly the others realised that the 'Six Brothers' had sunk, and all boats made at once for the spot to help the crew. With great relief, once it was certain that all hands were safe the fleet returned to Falmouth and so ended racing for that day. None of the 'Six Brothers' crew suffered any after effects and were all up at the Seven Stars Pub in Penryn that evening to join in the celebrations after the days racing with George Folkes, who had put up the cup.

Next day Dougie Rowe, the diver who raised 'Winnie' was contacted and he set about raising the 'Six Brothers'. As she was in deep water air bags were necessary and after a few problems, not least of which was the strap holding the air bags bursting and the bags shooting out of the water from 100ft. like Polaris rockets. She was brought to the surface, virtually without damage and after a lot of hard work by her crew was racing again one week later.

Though 'Winnie' and the 'Six Brothers' were the only boats ever to have sunk racing, it is not uncommon for a boat to fill up enough to have to stop and bail out with perhaps the floorboards afloat. One or two boats have sunk whilst dredging though, one of them being the 'Happy'.

Back in commission again none the worse for wear 'Six Brothers' well reefed for bad weather.

Rivalry amongst the boats is keen but friendly, protests were almost unheard of though there have been several in the last few years. The reason for this is that boats are raced in the spirit rather than the letter of the law and even if a boat has the right of way, the skipper will always give way rather than risk a collision, as the boats are too precious for their owners to risk any damage.

With their vast sail areas, reminiscent of the pre-war big yachts the working boats commanded the same sort of following ashore. Racing used to be keenly watched from one of the many vantage points ashore and afloat, each spectator willing their favourite on to victory, though interest here too has waned over the last few years.

Each port has its own particular favourite, for Falmouth, Toby West's 'Victory', for Flushing – the Laity Family's 'G. Glasson', for Mylor – G. Vinnicombe's 'Six Brothers' and for St. Mawes – it used to be the 'Florence' sailed by J. Sawle, until she was returned to her owner, Tom Burley. There have, of course, been many other favourites in the past, but these are the main ones in recent years.

Large crews are needed to race the boats as the gear is heavy and so an average crew would be about eight on a scratch boat and five on a smaller handicap boat. On a scratch boat two men would be needed to work the jib, two on the staysail, one on the mainsheet (two or more in heavy winds), one working the backstays and one on the helm.

The boats have always been divided for racing into two classes, over 25 ft. and under 25ft. Until 1975 both raced together with a separate cup being given for under 25 ft boats, when one was offered, otherwise if there was only one cup it would be presented on a handicap basis.

A few of the many trophies Toby West has ammassed over his racing career.

The handicaps are worked out annually by the skippers and crews and are kept for the season. After the 1974 season with an increase in the number if scratch boats racing, it was felt that there were enough handicap boats to make up a good class on their own. The same applied to the scratch boats, so they split up and with the handicap boats starting five minutes earlier there would be a better chance of all the boats finishing about the same time instead of the faster scratch boats a long way ahead. Many cups were donated to give the handicap class its own to race for, as the majority of the existing cups had been donated to the scratch boats and so had to remain with them. The scratch boat class is open to any Working Boat irrespective of size but it must race without a handicap, and conversely, any boat may race in the handicap class. Thus, if a 28ft. boat owner felt he could not afford to kit his boat out to a standard that would give him a chance to compete against the other scratch boats, he could still race in the handicap class.

Another reason for deciding upon the two classes was it was felt that a slower handicap boat racing from the back of the fleet could watch the leading scratch boats mistakes and it would be an easy matter to avoid them and thus save time against the leaders and gain an unfair advantage.

The only other difference between the two classes is that whilst the Handicap Class was allowed to retain spinnakers and at the time of the split in 1974 it was agreed that the scratch boats should not. The reasoning behind this particular decision was that it was felt by some of the skippers that as few owners actually owned their own light terylene balloon spinnakers, those that could afford them or had friends with yachts who would loan one to them, would have an unfair advantage as such a sail could give a boat an unassailable lead off the wind after long hard close racing on the wind. This was not received with great

Balloon spinnakers like this one on 'Katherina' were felt to be against the best interest of the class.

enthusiasm by all the crews, as downwind sailing with spinnakers has always been the most exciting part of their race, however, it seems to have worked well to date and demands for spinnakers to be reinstated on the scratch boats have so far been outvoted. The rules and regulations are made by the Falmouth Working Boats Association, which consists of a committee elected by the skippers, all the crews participate in the discussions, but the final voting is left to the skippers. The purpose of the Association is to safeguard the interests of the fleet and to raise the money needed for the insurance of the racing fleet, the cost of which is high and is raised by profits from the Annual Dinner and Dance and usually a 'get together' during the winter. Then in 1979 it was decided to get properly organised and the Association was formed and sold membership to help raise the cost of insurance and regulate the class. A handbook is now published and that too has helped raise money, so for the first time in its history the Association finds itself with money in hand.

This perfect suit of sails on Paul Williams boat 'Serica' shows what can be achieved with a gaft cutter.

Naturally, the sails used for racing have kept up with the new developments in the yacht racing world, and the days of going around with working gear are long gone if a skipper wants to compete seriously. The sails on the top boats are now of a very high standard and in some cases have got as near to perfection as is possible. Winches too are now the order of the day and the display of winches on board 'Hammer' Laity's 'Helen Mary' is an impressive sight. Sail areas too have increased dramatically and the forerunner in this development was Toby West in the 'Victory'. Other boats soon followed and the optimum sail area was found and adhered to. The only boat to go beyond this again as regards area was the 'Florence' whilst John Sawle raced her and her rig was larger than any boat had put on before or since. Her spars are of huge proportions for a 28ft. boat as her mast is 44ft. 6 inches, the bowsprit 26ft. overall and 15ft. outboard, the boom 22ft. overhanging the transom by 6ft. and gaff 19ft. Many argued that she was no longer a true working boat but a yacht, as she would not be able to work with this massive rig. However, she started dredging again in 1979 (the first time for twenty-five years) and the big gear proved no problem even though she

'Helen Mary' sailing hard with a couple of rolls in the main.

was worked single handed without an engine. The trend at the moment is to concentrate on improvements to cut and shape sail, slots, etc., and some boats use even Cunningham holes and many other boats of string to play with in order to alter sail shape during the race. Another recent development has been the advent of small staysails. Some skippers are finding that the smaller staysails will allow the boats to point closer to the wind than the larger traditional balloon foresails and the increased efficiency outweighing the loss in sail area. With all these developments it was felt by the Association that some sort of rules should be laid down to arrest the progress a little, for a situation might develop whereby many skippers might be forced out of racing by the spiralling costs involved trying to keep up with the new sails and gear needed to compete, if one boat were to get too far in advance of the rest. Obviously if one boat were to be sponsored by somebody with plenty of money and cheque book racing hit the class, it would not be desirable. In order to try and forestall this situation the Association drew the following rules, amongst others, in the hope it would stem the tide for a while.

1. No moveable lead ballast exceeding ½ ton.
2. No lead external ballast keels.
3. No encapsulated lead keels.
4. No external iron ballast keels over 2 tons in weight.
5. No encapsulated iron keels over 2 tons in weight.
6. No aluminium masts.
7. No more than four winches.
8. No new sails within three years.

The reason for the ban on lead keels and over ½ ton internal lead is the high cost of lead. Money could win races for the few who could afford to put most lead in their boat, as lead is the best form of ballast as it can be put lower in the boat due to its density. Another problem that has arisen is that with fibreglass boats coming into use, the traditional fixed ballast keel is disappearing, as it is easier to mould the shape of the keel into the hull and fill it inside with ballast, thus the internal and external ballast on a conventional wood craft become merged into one inside the boat. Obviously this has advantages for the builder as well as improving the stability and speed of the boat, and is desirable in every respect except unfortunately it does give them an advantage over the older wooden boats. It is a problem without a real solution and it has to be hoped that no boat gains too big an advantage from this development.

The two ton limit was agreed as most of the older wooden boats carry an iron keel of about 1.5 tons, and it would have been too great a strain for them (not to mention the expense) if they were to increase them by much more. By restricting the new boats to a maximum of two tons they only have a small advantage from their external keel as opposed to a big one!

'Victory'

Alloy masts would provide such an advantage that should one be allowed to use one racing the entire fleet would have no option but to follow suit in order to remain competitive.

Winches are very expensive pieces of equipment and most fishermen are unable to afford new ones and so have to buy the yachtsmans 'cast offs', and it was decided to put a limit on the number allowed in an attempt to reduce the advantage gained by the wealthier working boat owners who now race with the fleet.

Quite clearly the rules were made with the intention of protecting the character of the fleet and minimising the effect of 'cheque book' racing, but did it really work? The answer is probably yes and no. It may well have stopped the complete free for all 'cheque book' racing, but it allowed a lesser one that changed the style of working boat racing enough to drive away most of the dredgermen. Its fatal flaw was that whilst it concentrated on ensuring that the 'right' boats were in the class, it failed to recognise the importance of having the 'right' men in them, men who would understand the way working boat men race and play, and not feel the need to change it. This as we now know did not happen, and the new men who joined the class by virtue of having a boat that met with the criteria laid down by the Association rules were able to project their own ideas about racing, and the social side for which the boats were renowned is now a thing of the past. Maybe all is not lost, for there are signs of cracks in the wall, discontent is rife and though no

Maybe not the most efficient shape, large headsails like the ones 'Elizabeth Mary' (above) and 'George Glasson (overleaf) are carrying, have gone out of favour with the modern working boat skipper.

George Glasson

obvious solution is in sight at least the enormity of the problem is at last being recognised. That can only be for the good as at least some effort may be put into seeking a solution, the problem won't solve itself!

A 'port of refuge' in these troubled waters has been the Silver Oyster Race, started in 1978 by Peter Grigg of Mylor. He conceived it as a race for dredging boats only on the dredgermens traditional day off, November 5th. Only boats holding dredging licences would be eligible to race, and dredging gear only the order of the day – no fancy racing sails to be pulled out of the bag to give an unfair advantage! It was an attempt to get back to the 'old working boat spirit' of racing which had, many felt, been lost during the summer racing. The prize, a beautiful trophy of a silver replica of an oyster, was certainly worth racing for, but the real prize was the days race and the prize giving in the Mylor Sailing Club afterwards. Most dredgermen felt that if 'they' wanted to take over our summer racing, 'they' could have it, but this day was ours and no-one could take it away from us. It was of course a great success and provided several years of racing at its most enjoyable, but when the disease hit the Fishery, dredging virtually ceased leaving no boats to compete. In an attempt to keep the race alive until the return of the Fishery's prosperity, the race was thrown open to non working boats. It is always easy to see with the benefit of hindsight, but it may well be that

Not a GRP boat in sight as the boats start of Flushing. Left to right 'George Glasson', 'Evelyn', 'Florence', 'Elizabeth Mary' (furthest away), 'Magdalena', the new 'Six Brothers' and 'Victory'.

a mistake was made in doing this as now the Fishery's fortunes seem to be reviving and the fleet slowly increasing again, there is evidence of a reluctance on the part of the fishermen to bring their boats out on Race Day. This is because the atmosphere of the early races has been lost to a certain degree, evidence yet again of keeping the men together not the boats. With luck though this should only be a temporary state of affairs as a few more boats on the Fishery should help to revive the enthusiasm that is lacking at the moment.

A measure of the fun that is had by all can be illustrated by an incident at St. Mawes Victory Cup race in 1979. The crews were all in the Victory Inn letting their hair down after an exceptionally hard race in strong winds. The cup was being filled and passed around, when a visitor watching the 'events' said he would like to fill the cup. He said that he had come down from London and was amazed to see so many people singing, drinking and having a good time together without it breaking out into trouble and that he would like to pay to fill the winners cup, to be passed round in the traditional manner. When Norman, the Landlord, pointed out that it was being filled with an expensive concoction of Champagne and Brandy he put £60 on the bar and instructed Norman to keep filling the cup until the £60 was spent. He did not attempt to use this gesture to ingratiate himself with 'the boys'

A nice straight line for this start, nearest the camera 'Victory', 'Serica', 'Winnie' (with 'Six Brothers' sails and spars), 'Elizabeth Mary', 'Florence' and 'Leila'.

but was happy to stay in the background and in fact, very few were aware of what had happened until told of it the next day. The working boats by virtue of their simple way of fishing without the pressures of modern life, seem to breed a type of person content with the simple pleasures of life, who work hard and play hard, fiercely independent, men who face risks often enough to accept and meet them calmly.

As in the cases of the 'Winnie' and 'Six Brothers', when boats are still lying on the sea bed in great danger of being seriously damaged, and men have narrowly escaped with their lives, they are still able to forget their worries for an evening and join the rest of the crews in the evening fun.

But nothing lasts for ever, and sadly this simple enjoyment of racing has all but disappeared in the wake of a new breed of working boat men. True, some of the 'old school' are left, but only one or two, and they're being swamped by the 'new school' and getting perilously close to being sunk.

The rot started in the late '70s when a number of factors combined to bring about what many consider to be the decline of working boat racing, in spite of the fact that a large fleet still race regularly.

Firstly several GRP vessels were fitted out to a standard beyond the pockets of most dredgermen and fishermen, giving the boats such an advantage of speed that even if they made several mistakes during a race (and they often did), a wooden boat sailing a perfect race would still finish well behind the GRP vessel. (The reasons for this speed advantage are mainly the way a GRP boat can be ballasted and are explained more fully in the final chapter). This procession repeated weekly with the same boats finishing up to a mile ahead had a demoralising effect and men started losing interest.

Secondly, the newcomers to the class were more interested in the racing than the 'social' side of racing, and a new phenomenon appeared, namely prizegivings at which the skipper and crew of the winning boat did not appear, which of course left the proceedings rather flat!

Thirdly, as men grew demoralised by the racing side as well as the social side they began to quietly drop out. The dredgermen were fast becoming outnumbered by the working boat 'enthusiasts', who were more interested in sailing in the fashion of modern yachts, that is with the emphasis on increasing boat speed by buying expensive gear and gadgets, and using a deep knowledge of the rule book to advantage. This is almost completely the opposite of the traditional working boat man's philosophy, who liked to win by beating the man rather than his boat beating the other boat. So the days of delaying the start for a boat that was late leaving the pub were over, and people had to get used to the previously unknown sight of protest flags fluttering from the rigging of working boats, and dredgermen had to learn to decipher commands hurled at them across the water as they rounded turning marks, which for all the significance they carried may just as well have been spoken in a foreign language.

Finally, the process accelerated, even more dredgermen dropped out as they were no longer getting any enjoyment from the social side and couldn't afford to compete with the 'cheque book' tactics of the new boat owners. So at last the situation has arrived where the takeover is virtually complete and the class is now run in the way the new men want and they should all be happy, except for the dredgermen who have lost their Saturday afternoon sport. But they are not all happy, as everyone can see that something unique has been lost, had the new boats been prepared to join the class and accept it as it was and not feel the need to 'improve' it, the 'good old days' would still be with us.

CHAPTER 5

The Oyster Fishery

Around the British coast in the early part of the twentieth century many oyster fisheries thrived. The most important of these were in the Blackwater, Colne and Kent (Whitstable) areas. Prior to this the oysters had been fished in deep sea beds, as well as these inshore beds, with boats working far out into the English Channel. The inshore beds were once of course worked by sailing smacks in exactly the same way as the Truro boats do still, but the difference was that by the end of the 1940's they had nearly all taken advantage of power and changed from sail. This spelt the end for the natural fisheries as no fishery can sustain unlimited dredging by motor boats, and as there was no regulating authority on these free fisheries they soon were finished. This combined with the severe winter of 1962/3 (which killed 95% of all oysters on the East Coast) is the reason that the Truro Oyster Fishery is the sole survivor. A sailing smack is a very inefficient tool for catching oyster for a variety of reasons. Firstly the dredges themselves have only an efficiency of about 15%, meaning they pick up only 15 out of every 100 oysters. Secondly they need wind and if there is no wind they cannot work, and in light wind can only work very slowly. Thirdly, oysters are not evenly spread over the sea bed, but lie in patches. As a sailboat has to drift it can only pass over a patch once and then has to sail back upwind of it for another drift before it can catch from the same patch. A sailboat would therefore make as long a drift as possible and try to pass over the known patches of oysters during the course of that drift. A motor boat once having found the patch can go round and round on it, fishing it until it has gone. Lastly, because of the inefficiency of the sailboat there comes a time when though there is still a large stock of oysters left on the bed it is not profitable to try and dredge them any longer. 15% when the oysters are dense is a good week's wage, but 15% after they have been thinned out by continuous dredging is not, so boats drop out one by one as they look for more lucrative employment. Fortunately for the oyster there are enough left behind to form a breeding stock and so the beds are replenished. In the case of motor boats the number of oysters left behind is far fewer due to their improved efficiency, and so the beds may not recover if stocks are depleted below a certain level. Once the stock drops below a certain level, the oyster seems to lose the will to propagate and then no spatful can be big enough to save the fishery. Surely there is a lesson to be learnt here and applied to other sorts of fishery?

In a good season the number of boats increases as happened in 1925 when the highest number on record of 50 sailboats and 87 punts was reached, but by 1935 the fishery had declined to such a low ebb that only 7 sailboats and 19 punts were licenced. So it can be seen when the Fishery is at a peak and can sustain a high level of fishing it does so, but when stocks are low and the danger of overfishing at its worst, the fleet contracts and gives the fishery the rest it so badly needs.

The restriction on every licence that "Boats propelled by Mechanical Power not Licenced", is the factor that has saved the Truro Oyster Fishery and preserved its fleet of sailboats. The dredgermen of the River Fal are only too aware of this fact and so will never put pressure on to change from sail to power.

Licence No. 751

CITY OF TRURO

TRURO OYSTER & MUSSEL FISHERY

Licence is hereby granted to Mr Davies
for the Fishing Smack or Boat called "Result"
2 Tons Register (to Dredge with 2 Dredge) Fish for,
and take with the said Boat, Oysters and Mussels from the ground situate within the Limits as described in "*THE TRURO PORT FISHERY ORDER, 1936*," from the date hereof, to the Thirty-first day of March next, subject to the Provisions of the "Sea Fisheries Act, 1868 (31 and 32 Victoria, Ch. 45) the Regulations of the said Order and the Bye-Laws (if any) from time to time made by the Corporation of Truro, under the powers for that purpose in the said Order contained."

Given under my hand this Twenty-Eighth day of September 1973

N.B.—The Ring to be used is not to be less than Two and Five-eighths inches in diameter on the inside thereof, and no Grapnel is to be placed further in than ordinary low water mark of Spring Tides.

Boats propelled by Mechanical Power not Licensed.

PK

TOWN CLERK.

(SEE BACK)

Oscar Blackford Ltd., Princes Street, Truro

CITY OF TRURO

Truro Oyster and Mussel Fishery

EXTRACTS FROM

BYE-LAWS AND REGULATIONS

1.—Close Season April, May, June, July, August and September.

2.—No Persons shall Dredge or Fish for or take any Oysters or Mussels between the hour of Three in the afternoon of one day and the hour of Nine in the morning of the following day.

3.—No Person shall Dredge or Fish for or take any Oysters or Mussels between the hour of One on Saturday afternoon and the hour of Nine on the following Monday morning.

4.—No Person shall remove any Culch or other material for the reception of spat. All Culch to be deposited in place from which it is taken up, and no Culch to be thrown or deposited at any point above ordinary low-water mark of Spring Tides.

5.—That the prohibition contained in Clause 9 of the Truro Port Fishery Order, 1936, for the removal by any person from the said Fishery or from one portion of the Fishery Grounds to another of Mussels of less size than two inches in length of any Oysters which will pass through a Circular Ring of such diameter as the Corporation shall from time to time direct shall extend to the Purchaser or Purchasers of any Mussels or Oysters as well as to the person or persons orginally dredging for or taking the same and that the Penalty on Conviction for Breaches of any of the Bye-laws or Regulations a sum not exceeding £20.

Copy of Dredging Licence

The Truro Oyster and Mussel Fishery Order 1876 is still basically the same as the present one and so can be quoted in order to show the extent of the Fishery. This particular Order was granted for a period of 60 years. It gives the total acreage of the Fishery as being 1743 and the boundaries as follows: Victoria Point (St. Clements), to Messack Point (St. Just), to Tarra Point (Mylor), to Sunset (Kea), to Victoria Point again. The Fishery is to be closed between May and August, only half the bed may be closed down at any one time, the ring size to be 2½ inches, and all profits to be put back into the Fishery. The cost of licences is set at 3s per reg. ton for the boat, 1s for the issue of the licence itself, 7s-6d per dredge in a punt and 10s in a sailboat. The prices for the Mussels are 1s per reg. ton, 5s per dredge in a punt, and 7s-6d in a sailboat. The cost of replacing a lost licence is set at 1s, and a fine of up to 10 pounds for anybody resisting arrest.

These licence fees are altered at the discretion of the Corporation, and have ranged over the years from 10 shillings per dredge in 1899 to the present fee of 100 pounds per dredge. This is the only revenue that the Truro Corporation receives from the Fishery, and this has to be used to run the Fishery and cultivate it when necessary. The records between 1887 and 1900 bear witness to the efforts that were made during that period:

1887: 40 pounds spent on young oysters to provide spawn, and 20 pounds allocated for the moving of cultch from one bed to another where it can be more beneficial to the fishery.

1890: £55 allocated for moving cultch.

1891: £25 allocated for moving cultch.

1892: 20 pounds allocated for moving cultch, and 30,000 oysters to be purchased at a cost of 25 shillings per thousand to provide spawn.

1895: Tiles were laid in an effort to catch spat, but as there are no subsequent reports on this project it must be assumed that it did not meet with a great deal of success.

The bailiffs worked for their money harder than usual during the period between 1887 and 1900, for there was trouble brewing which culminated in the court case that we still celebrate on the Fishery every year by not dredging on November 5th, the day the dredgermen were victorious. It must have been a difficult situation for the bailiffs as many of the dredgermen were friends or relations, and it was they who were expected to enforce a very unpopular demand by the Corporation. The dredgermen had always laid their oysters on layings on the foreshore and now the Corporation decided to claim them as their own and impose a charge on the dredgermen for using them. Predictably this was met with strong opposition by the dredgermen, who eventually banded together in 1900 and hired a Queens Council to put their case, and won the day. The following entries from the minutes of Truro Corporation committee meetings, though brief, do manage to convey something of the atmosphere of the dispute.

1887. Bailiffs instructed to make a list of beds, and a charge of 1/- be made.

1888. That any person depositing cultch on the Fishery on the foreshore within the limits of the Fishery shall be liable to a penalty not exceeding 5 pounds.
Merchants and dredgermen to be charged 1/- for a 20′ run on shore for laying.

1889. Small oysters found on beds and steps taken to prevent it happening.
In order to see if The Order is any good, issue a summons against William Henry Gunn for having illegal oysters in his possession.
Withdraw prosecution due to difficulties in obtaining a conviction.
Mr. Burley complained oysters laid by him in Coombe had been taken or destroyed by the Water Bailiff and claimed 20 pounds for the loss sustained.

1892. 400 small oysters found on one bed in Mylor and removed and put back on the Fishery.

1893. Stakes had been placed on the Fishery to create beds and the Bailiff removed them.

1894. More stakes put by fishermen.
Water Bailiffs boat 'stolen, sunk, or otherwise illegally disposed of'. Reward of 10 pounds for information.

1899. Debate on fishermens foreshore rights.

1901. Council wish to appeal against decision of Mr. Justice Wills in action against Rowe and Kemp.

1902. It is decided to let the appeal drop.

And so dredgermen since that day have Guy Fawkes Day, November 5th as a holiday to celebrate and remember this famous and very important victory.

The Truro Oyster Fishery

The fishery is owned by the Corporation of Truro and called the Truro Oyster and Mussel Fishery. It is now taken over by the new government department of Carrick under the re-shuffle but still managed by the Truro Harbour Master as before.

Every dredgerman pays a licence fee for each dredge used, the norm being 2 to each man, and this is the only revenue the regulating body receives. From this they have to finance the policing, administration and preservation of the fishery. The policing is done by the Oyster Bailiff. He has a motor boat and leaves Truro every morning to be on the Fishery for 9 am when dredging commences. He has a wide area to cover and naturally it is left to his discretion as to how best to do his job. As he passes down the river he will pass all the punts working in the upper reaches of the Fal. Whether he may stop and chat or look through some of the catches is dependent upon his whim. There is far more to the job than just checking oysters for size and making sure the boats stop and start on time and so a few words of conversation or a stop for a bit of gossip will help him keep in touch with the men. One bailiff, for example, Guy Saunders, was only too happy to pass messages from one boat to another and even brought out biscuits to feed the dogs on board the various boats every day, and so performed a social service as well as his other duties. It is this sort of attitude that makes the fishermen feel they are part of a big family.

The administration involved is not vast and the bulk of the work is done by the bailiff, though it has become increasingly more time consuming. With the risk of pollution due to the increase in the volume of commercial traffic and the threat of invasion by the dreaded Japanese seaweed, and the fight against Bonamia, more time has to be spent in preventing damage to the fishery. There are of course other duties such as printing and issuing licences and the upkeep of the bailiff

The popular oyster bailiff Guy Saunders indulging in his passion, gondolas. Here seen rowing his own in the upper reaches of the Fal.

and his boat, but except on rare occasions, there is little else. One such occasion though was in 1969 when the Truro Corporation purchased a large amount of queen shells (a queen is rather like a small scallop), as the cultch on part of the fishery was starting to thin out. The 'cultch' is the layer of shells which are vital to the continuation of the fishery as they are needed for the young oysters to settle on (see chapter 6). Normally the cultch replaces itself naturally as the many oysters that die of natural causes provide the layer of cultch over the beds, but sometimes it gets a little too thin in places, as happened on this occasion, and so Truro Corporation decided to give nature a hand. The Corporation provided the shell, purchased from Plymouth and the men transported it down to the fishery from Truro in their own boats and scattered it where it was required. This was done in conjunction with the scientists of the Ministry of Agriculture and Fisheries from Burnham who keep a regular eye on the Fishery on our behalf and advise on any action they may feel is necessary to the Corporation. This example illustrates well the co-operation that exists between all sides involved in the Oyster Fishery.

As far as protection goes, that is a problem that is becoming greater than ever before. The biggest expense facing the Corporation is the need to combat the Japanese seaweed when it reaches Falmouth. This fast growing seaweed could smother the fishery unless a boat is employed full time during the close season in the summer when it grows its fastest. As this would be a new experience for all concerned (the weed has only recently reached this country) research needs to be done on the various methods of combatting it, and this of course requires time and money. However, this is a challenge that has to be met, and the Corporation are facing up to it.

The Fishery is located on the River Fal upriver of an imaginary line drawn between two upright granite posts, one situated on Pennarrow Point and the other on Messack Point. The area worked by the sailboats is the Carrick Roads section of the Fishery, whilst the punts work the more sheltered and narrow sections of the fishery where sailboats would find it too confining and the wind would be too flukey as the banks of the river are high and covered in trees. Ideally a sailboat needs a large open area to work and this it finds on the bank in the Roads. There are three banks and these are called East Bank, Mylor Bank and Parsons Bank. The East Bank is on the eastern shore of the river, the Mylor Bank is off Mylor and Parsons Bank off Loe Beach. The border line between Messack Point and Pennarrow Point was on the boundary line of the old port of Truro and below that came under the Authority of Falmouth, though now it is all one under Carrick. There were of course oysters on the Fal side of the boundary too, but they had their own licensing

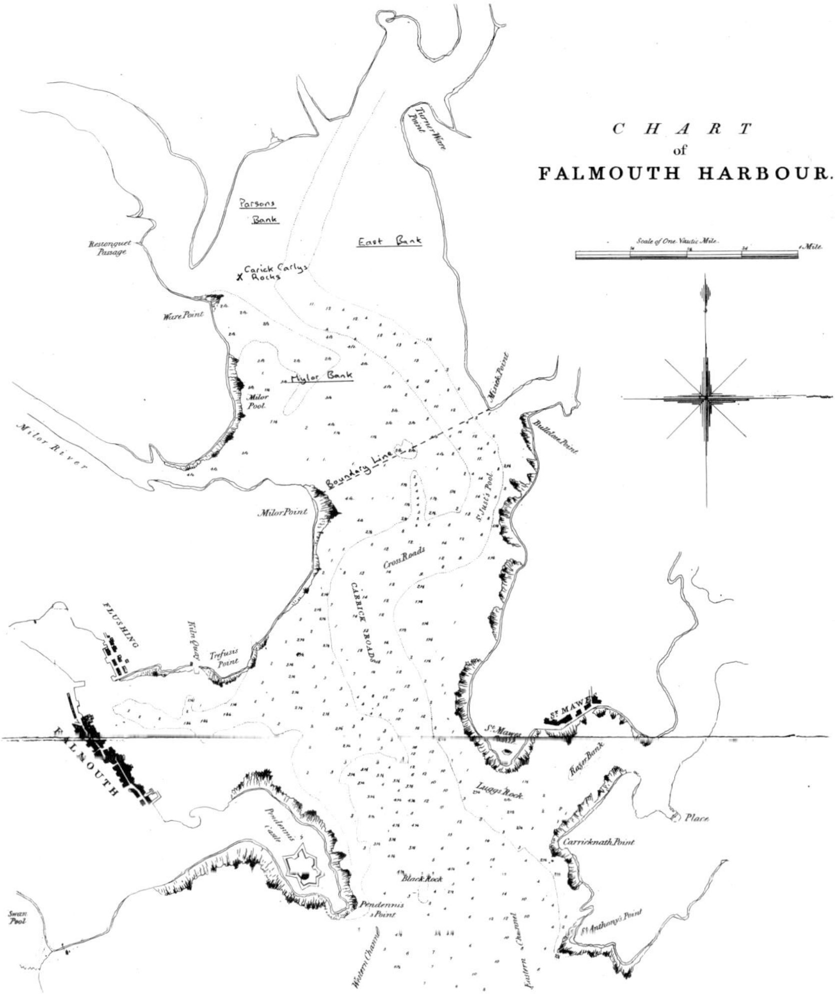

system as did Truro.

The Falmouth Oyster Fishery

Below the Truro Fishery boundary began the Falmouth Fishery, which was run the same way as the Truro Fishery, except by Falmouth Town Corporation as opposed to Truro Town Corporation. There are few records of this Fishery left, so it is difficult to present an accurate history of it, but the original Falmouth Fishery Order of 1877 is still preserved and so we have some information on it. It was granted to the Borough of Falmouth for 60 years and this would appear to coincide with the time that the Fishery closed down in 1937. The boundary of the Fishery was as follows: Messack to Tarra Point, Castle Point (St Just) to Carricknath Point, to Zone Point, Pendennis, a spot on west side of Bowyers Cellars, across to Sailors Creek and back to Tarra. The Fishery Order further stated that the beds were not to be closed for more than 2 years and up to half the bed only may be closed at any one time. Licence fees at this time were 1 shilling each and a charge of 3 shillings per ton was levied on each vessel. On top of this, the cost for each dredge used was 7s-6d per dredge in a punt, and 10 shillings per dredge in a sailboat. Provision is also made in the Order for Mussel dredging, and the costs here were 1 shilling per ton and 5 shillings per dredge for a punt, 7s-6d for a sailboat. The ring size could vary between 2″ and 2½″ (at the discretion of the Borough) for oysters, and 2″ for mussels. Any wrong-doer who resisted arrest was liable to a fine of up to 10 pounds, though there are no references in the minutes of the committee meetings of this ever happening. Finally, the Order stipulated that all profits must be put back into the Fishery. The two Fisheries of Falmouth and Truro seem to have worked well with each other, and indeed there is no reason why they should not, for they were not in competition with each other. In fact the two Corporations worked quite closely together and in 1889 it was suggested that they 'co-operate for protection of the Fishery's', and it was reported that this idea was favourably entertained. Again in 1892 the same theme recurs in the Truro Corporation's minutes, where it is recorded that the Falmouth Oyster Bailiff is to be appointed for Truro also. The dredgermen normally paid to dredge in both Fisheries, working the Truro beds on the spring tides and Falmouth on the reaps, but as the Falmouth Fishery seems to have been in a slow steady decline from the early 1900's onwards, presumably dredgermen became less inclined to spend money on licence fees at Falmouth. A practice that seems to have been peculiar to the Falmouth Fishery was that of hoisting a flag on board the bailiffs boat to warn the men when the day was drawing to a close. Five minutes before the end of the day the flag would be hoisted half way up the mast, and then right to the top when it was time!! The Falmouth Fishery is still dredged by a few punts in the Penryn River area, and there has been a little bit of activity on the North Bank again over the last few years, but it is highly unlikely that the Fishery will ever be worked properly again. There is no longer any Regulating Authority in charge of it and so the boats dredging on it do not have to pay any licence fee or work any set hours, or even any season, though in practice a gentlemen's agreement is kept and the boats work the same hours as the others on the Truro Fishery.

Year	Sailing Boats	Punts	Quantity	Value(£)
1910-11	21	74	–	–
1911-12	31	65	–	–
1912-13				
1913-14	33	115	–	–
1914-15	15	65	–	–
1915-16	14	50	–	–
1916-17	17	49	–	–
1917-18	17	43	–	–
1918-19	32	69	–	–
1919-20	33	101	–	–
1920-21	39	95	–	–
1921-22	37	129	–	–
1922-23	45	144	–	–
1923-24	30	153	–	–
1924-25	50	187	–	–
1925-26	39	145	–	–
1926-27	22	82	–	–
1927-28	14	58	–	–
1928-29	22	91	–	–
1929-30	23	73	–	–
1930-31	9	58	–	–
1931-32	15	50	400,000	1400
1932-33	14	49	375,000	1312
1933-34	10	43	225,787	787
1934-35	11	29	191,000	669
1935-36	7	19	103,000	260
1936-37	11	35	341,000	853
1937-38	18	41	–	–
1938-39	14	47	–	–
1939-40	9	25		
1940-41	13	42	–	–
1941-42	15	42	–	–
1942-43	13	39	–	–
1943-44	17	51	–	–
1944-45	19	45	1,116,000	8,649
1945-46	24	63	1,732,000	13,856
1946-47	26	69	1,519,000	7,345
1947-48	23	67	1,782,000	8,910
1948-49	15	56	869,600	4,348
1949-50	20	49	1,543,800	5,403
1950-51	14	47	1,815,000	5,445

1951-52	19	39	2,775,800	11,100
1952-53	21	40	3,500,000	13,250
1953-54	19	47	3,502,000	15,000
1954-55	21	54	3,232,000	14,710
1955-56	26	91	1,564,000	11,162
1956-57	20	77	1,811,000	12,272
1957-58	23	66	1,820,000	14,650
1958-59	21	51	2,010,000	12,400
1959-60	21	66	2,000,000	12,250
1960-61	13	50	484,000	3,200
1961-62	10	–	–	–
1962-63	10	–	–	–
1963-64	20	–	–	–
1964-65	16	–	–	–
1965-66	25	–	–	–
1966-67	25	–	–	–
1967-68	30	–	–	–
1968-69	30	–	–	–
1969-70	35	–	–	–
1970-71	29	–	–	–
1971-72	22	–	–	–
1972-73	18	–	–	–
1973-74	No further records			

This table is of interest as it shows the fluctuations in the size of the sailboat and winkboat fleets between 1910 and 1973. It was obtained through the records of Truro Corporation and though the values will be an accurate one for the quantities of oysters, the quantities themselves probably do not give the true totals, merely an educated guess on the part of the bailiff.

CHAPTER 6

The Fate of the Oyster

So far we have neglected the oyster itself, which is the reason for the fishery and the fleet.

The oysters on the Truro Oyster Fishery are of the only true British species, that is the European Flat Oyster - Ostrae Edulis or Edible Oyster. It grows throughout the British Isles but must not be confused with the 'Native' oyster. The 'Native' is the same oyster as the European flat oyster for the name Native refers only to the area of origin. In order to qualify for the distinction of being a Native the oyster must have been reared between South Foreland and Orford Ness, where the bottom is all of London Clay. An oyster may be relaid in that area, say for example in the Whitstable area having begun life in Cornwall. After spending some time in the water it can then be sold as a Whitstable oyster, but not a Whitstable native. Neither can an oyster be called a Falmouth Native as the Fal is outside the qualifying area. The oysters breed in the summer months during the close season, hence the rule of never eating one when there is no 'R' in the month. As far as the private companies are concerned they may sell during the months of September and April but the fishery is closed then to give the oysters a better chance to breed. During the summer, as soon as the water starts to warm, the oysters prepare to spawn, and normally this occurs when the water temperature has risen to 15°C. The female fertilises the eggs within her shell and if any oyster is preparing to spawn a milky substance is apparent inside. These are the eggs and they go through three stages prior to expulsion

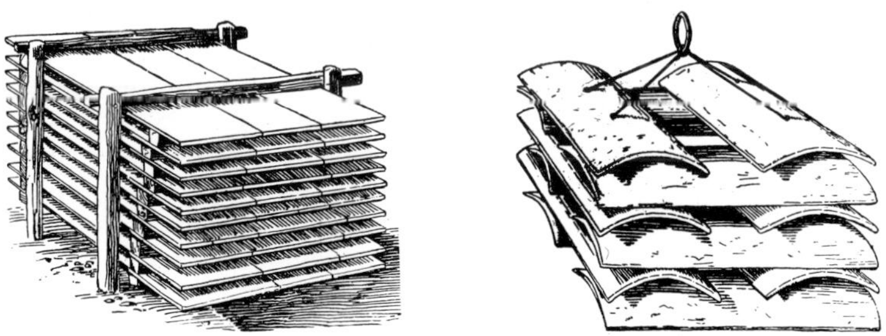

Spat collector such as these made up of tiles are commonplace throughout France where spat collection is a large part of their oyster industry.

from the female. The first is when the eggs appear a milky white colour known as 'Whitesick', the second is when they have turned grey and then the oyster is termed 'Greysick'; lastly when the eggs are black then the oyster is 'Blacksick' and ready to expel the eggs into the sea. In France where oyster cultivation is a major industry, every one of these stages is closely followed, for as soon as the Blacksick stage is reached, traps are put out to encourage the spat (as the settled larvae is called) to settle and thus the following years stock is collected.

Once expelled from the parent oyster, the young have a free swimming larval stage at the end of which they must find a fairly clean piece of shell to which they attach themselves. The longer spent swimming and searching for a place to settle, the greater the chances of it being eaten by a predator (such as a fish) and so it must settle quickly. Only about one in 10,000 will survive long enough to settle and as an oyster may liberate 1 million eggs each summer, it can be seen that about 100 may reach the settlement stage. Having settled, the young oyster now starts to form a shell and is known as spat. Here the spat will stay, adding layers of shell or 'shoots'on the outer edge to increase in size until it is about two years old. Now it becomes known as brood and is about halfway through its career. Many will have died by now as they are susceptible to predators, the worst on the Fal beds being the Oyster Drill or Rough Tingle. This mollusc can drill straight through the shell of the oyster and then scoops the flesh out of the oyster whilst it is still alive. Starfish are partial to oysters as well as us, and kill them by wrapping their arms around the shells of the oyster and applying a steady pull with the suction pads on the arms until the oyster starts to weaken and open slightly. As soon as this happens the starfish extrudes its stomach through the mouth, between the oyster shells and the digest action that follows causes the oysters muscles to relax and the starfish can settle down to its dinner. Starfish particularly like young oysters as they are easier to open, but fortunately we do not have too many starfish on the Truro fishery. Other predators include crabs and bristleworms, the boring sponge 'cliona celata' more commonly known as 'kanker'. Oysters with the bristle worm 'polydora' and the boring sponge must be removed from the fishery at once as they both spread rapidly and destroy the shell of the oyster making it unsaleable in the early stages of attack, but finally killing the oyster when the shell becomes too badly infested. In addition to the predators, the competitors for the same food as the oyster needs have to be culled. These include 'slipper limpets' which can take over a fishery completely if allowed to do so, smothering the oysters making their control vital. Once they become brood they have to face their biggest enemy, who is of course the dredgerman.

Once caught and failing to go through the oyster ring, the oyster is put into a bag and taken ashore. There are a variety of merchants wishing to purchase these young brood oysters, and the dredgerman may choose whom he prefers. If the oysters are sold immediately to

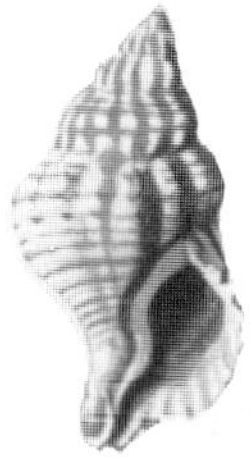

c Urosalpinx cinerea
OYSTER DRILL
× 1

Carcinus maenas: SHORE CRAB × 1-½

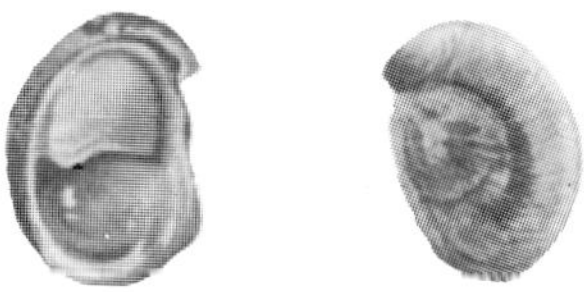

Berthella plumula LIMPET

Asterias rubens
COMMON STARFISH
× 1-1/6

Marthasterias glacialis: SPINY STARFISH × *c.* ½-¼

the merchants, they are sold as 'Freshcaught' oysters, though some men prefer to lay the oysters back on their own private beds and sell them after the season has closed as 'Bed' oysters for they can now command a higher price. Not all choose to do this for if the weather should become too cold, as happens frequently on the east coast, the oysters may suffer a high mortality rate.

Oyster merchants 'Seasalter and Ham Oyster Company' and Harry Johns Oyster Company of Flushing receive a shipment of brood oysters from Brittany for relaying on their oyster beds in the Penryn River. These will supplement stock purchased from the Truro Oyster Fishery.

Having purchased his oysters, the merchant relays them on fattening beds for at least one more year before they are large enough for the restaurants. During the time spent on the beds a loss of one third is incurred during each summer spent there. This loss is caused by spawning which weakens the oysters considerably and is one reason for their high cost in restaurants. Other contributory factors towards this massive mortality are 'dredge knock'. This is quite simply what it says, for the dredges are heavy and can kill a proportion that they pass over. The oysters are taken ashore now and put into purification tanks where they must remain for 36 hours to kill all the bacteria. They are now ready for grading and packing and to be sent to market. Five grades are usual, grades one, two, three and four and finally the smallest grade called buttons, and naturally the price increases with size. The oysters journey finally ends on a plate in a restaurant.

CHAPTER 7

Working Under Sail

DREDGING OFF PARSONS BANK

Although the boats from different areas vary a little in design, they only vary within acceptable limits and these limits are defined by the fishery itself. The boats length is ideally between 22′ and 30′. as any smaller would be too cramped for one man to work in comfort. Having said that however there have been a few boats as small as 18′. such as the 'Thelma', 'Magpie' and 'Flying Cloud'; anything over about 30′. is usually too deep and starts getting a bit of a handful for two men in bad weather. The very small boats did not last very long as their owners all bought larger boats as soon as they could. Mike Parson's little 'Thelma' broke up on Greatwood Quay during a gale in 1962 and he replaced her with the 24′. 'Leila' which he built himself. The other dredgermen had a 'round robin' and raised enough money to buy the materials for him. Kenny Cork purchased the 27′. 'Evelyn' after he sold the 'Magpie 2', though she is still sailing as a yacht. On the top end of the scale

The little 'Magpie', not to be confused with the restricted 18 footer 'Magpie' that Ken Corke's father George raced for many years.

A proud young Kenny Corke displays the first oysters caught in his career at the end of the first day out in 'Magpie 2'.

Nearly every boat seems to have a dog on board, here Jet is seen on board the 'George Glasson' with Paul Laity.

both 'Morning Star' 34′. and 'Lily' 35′. have worked on the fishery in recent years, and they both have been handled as well as their smaller counterparts, most dredgermen opt for a slightly smaller boat out of choice. The draught is dictated also by the fishery as the optimum draught for working at low tide is about areas of the grounds that cannot be worked at low tide with the extra draught. Too little draught and the sailing abilities of the boat would be adversely affected. Beam varies between 7′ 6″ and 10′ on a boat of about 27′ The boat with 10′. beam will obviously have more room and it will also be a more comfortable boat to work on as it will heel less sailing and probably sail faster too. With regard to the rig, this too is controlled by the nature of the work. A gaff rig is chosen for 2 main reasons. The first is that it is probably the only rig that allows the boat to be controlled sufficiently accurately to perform what is expected of it, and the second is that it generates sufficient power to drive the rather bulky hulls and still have a short mast. The average mast on a 28′ boat being about 38′ but if it were to be Bermudan rigged it would need to be half as high again. I think we can safely assume that if there were a better rig for the job it would have found its way into

The narrow beam of 'Result' made her rather less comfortable than would be liked for dredging purposes, as soon as the wind increased she would be permanently 'on her ear' making it a constant battle for a foothold.

From left 'Vera', 'Muriel' and 'Mayflower' at work.

use by now. Finally the long bowsprit which is about 12′ outboard, gives the jib far greater leverage and subsequently more control over the boat results. Whilst they are working the boats look quite different to the way most people who see them in the summer would expect. This is because the amount of canvas used is far less than during racing. For instance, the smallest jib used for racing (i.e. in bad weather) would be about the size of jib used only in very light weather when dredging. Probably two reefs would be put in the mainsail for dredging on a day when full main would be carried racing. The working staysail is about one third the size of a racing sail and no topsail is carried. A storm jib, or spitfire, normally used dredging might measure 15′. on the luff, and so clearly everything is set up for ease of handling.

With only a few exceptions ('Boy Willie' and 'Twilight' are two) all the boats are open decked to the mast to give a forecastle for storing sails etc. on narrow waterways about 18″ wide upon which the trays rest, and the dredges are emptied. They are very important when sailing too, as when the boats are hard pressed it prevents the majority of the water from coming into the boat. The only reason the bigger boats are open as opposed to flush-decked is that it would require constant bending to have to cull the oysters from deck level. The two boats that have had flush decks in recent years have had bulwarks high enough to support trays at waist level as they are 33′ and 36′ in length.

In order to dredge oysters from one of the three banks on the fishery there are two factors that must be correct, and one is not sufficient if the other is wrong. The first is the tide which determines which direction you must work a bank, and the wind determines if you can or not. All boats must dredge with the tide because when they are dredging they do not sail but drift, and the sails are used to keep them at about right angles to the tide, though when there is insufficient tide the boat can be sailed forward as well as drifting in order to cover more ground, but not too fast as the dredges would lift off the bottom.

This sideways journey with wind and tide down a bank is known as a 'drift' and then each drift is completed (i.e. the boat reaches the limit of the bank it is working) the boat has to sail back to the starting place and begin again. On an average day a boat would probably complete six to eight drifts. There are many skills necessary to make a good oyster dredgerman; being able to sail, intimate knowledge of the oyster beds and tides, how to make and use dredges etc. but probably the most important thing is being able to make a boat drift at exactly the right speed in the right direction. As conditions vary from day to day and sometimes from drift to drift, the combinations of sail required to make the boat do exactly what is required of it are infinitely variable. The course pursued by the boat is vital and a few yards off course may result in missing a patch of oysters, going off the edge of the bed and

Boats dredging have a totally different beauty compared to when they're racing. 'Pet' on the left, and 'Leila' on the right.

into the deep water of the channel where the dredges would not touch the bottom, or fouling a hitch on the bed. At its simplest this is achieved by lowering the staysail, backing the jib to windward, paying off the mainsail and lashing the helm to leeward. Set up in such a manner a boat would drift down tide and down wind, though across the wind in a forward direction as well. When the mainsail is paid off the forward drive is removed, but when the jib is backed to windward it pushes the bow around until the mainsail starts to fill again. When this happens the boat starts to move slowly forwards and as soon as it does this the lashed helm pushes the boat back up head to wind and the process begins again. By lowering the jib partially and therefore reducing its area, it would take longer to push the bow back and fill the mainsail and therefore less forward motion would occur if this were found to be necessary. Another common sight is with the gaff partially or completely lowered by the peak which has the effect also of reducing forward motion, as there is less sail working to push the boat forward. Thus it can be seen that the variations are endless, often sail being altered several times in one drift in order to speed the boat or slow her, to change direction in order to drift over a good patch of oyster or avoid another boat. But whatever the reason it must be done in plenty of time for if you find yourself somewhere you would prefer not to be, by the time the sails are hoisted and dredges hauled you could be on the rocks or entangled with

Each sail configuration is carefully planned and performs in it's own way. 'Evelyn' (left), 'Beatrice' (right).

A glance at 'Evelyn' on the left might give the impression something is wrong, but when you study 'Pet' and see her sails are arranged in an identical way it is clear that they are this way for a purpose.

another boat. It may sound a fairly simple matter to drift a boat down wind, but I wonder how many experienced yachtsmen if cast off their moorings, told to throw four heavy metal dredges over the side and sail their yacht under control to a specified spot down wind, would actually be able to do so. It is no reflection upon them that they could not, as a yachtsman has learnt only to sail his boat, whereas dredging undersail is a totally different skill.

In this manner the boat drifts down wind by itself leaving the crew free to work the dredges. It is usual for one man to work two dredges so a two-handed boat would work four. The dredges are triangular about three feet long and three feet wide and are quite heavy (about 30 lbs) as they have to stay on the bed even if the boat is jumping about on the surface in a big sea. They are towed by ropes evenly spaced along the windward side of the gunwale. (The knife on the back edge scooping up the oysters and guiding them into the net attached to the dredge.) After the dredges have been towed for a few minutes the first is hauled and emptied into the tray on the side deck. The assortment of oysters, shells, crabs, stones etc. (known collectively as cultch) is quickly sorted. The oysters are picked out, the predators (crabs, starfish, oyster drills) are killed and then the cultch is tipped back over the side. This is important as the newly born oyster, known as spat, need plenty of cultch to attach

Kenny Corke manages to squeeze in a few moments for a 'fag' as he decided upon his next dredge!

themselves to otherwise they drift out to sea to be eaten by fish whilst still in their larval stage. The next dredge is then hauled and the process repeated until the end of the drift, then the dredges are hauled aboard, sails set and the boat goes back upwind to begin again for the next drift. As the dredges are hauled continuously there is no time to stop even for a cup of tea during the drift, and in bad weather the single handers may not even get time on the way back for the next drift as the sail back may be as hard work as the drift itself. Once the oysters are caught they have to be measured by means of the ring issued by the bailiff which is 2⅜″ diameter, and any that can fall through must be returned to the fishery

Johnny Walls chats with Clinton Powell on 'Gypsy', as he sails 'Beatrice' back across the drift.

in order to maintain a breeding stock. The average age of the oysters that are legally acceptable is about 2 years, it will have started life on the fishery as all the stock is self-propagating and will then be sold to an oyster merchant. These buyers will have to relay them on their own private fattening beds for at least another year during which time they will suffer a mortality rate of about 30%. Finally they will be dredged up again, purified under ultra violet lights and from there will be sent to the hotels and restaurants.

There is another fleet of smaller boats which work the Fal Oyster Fishery and these are known as the 'Winkboats'. A winkboat is a punt about 15′ long which is used for dredging by sailboatmen when it is too calm for sailing, as well as by others who prefer this method to sailing.

A wink boat.

Its chief advantage over sailing is that once a patch of oysters is found it can be worked far more efficiently.

The technique used is to tow a dredge along the bottom behind the punt by use of a hand winch, called a 'Wink' hence the name Winkboat. An anchor is first thrown over the side and the dog is slipped off the winch so the boat can be rowed stern first down tide until all the line, usually about 50 fathoms, has been paid out. The dredgermen stand up to row, and the name given to this rowing position is 'sheaving'. The wink is in two parts, one to operate the headline and one for the dredge warp, and so once the end of the headline has been reached the dog is released on the part that controls the dredge warp, and the dredge can then be tipped over the side. When the correct amount of dredge warp has been paid out the wink is locked so the dredge warp stays at a constant length, and then the boat is wound back upstream to the anchor on the winch. Once over the anchor the headline is locked on the winch, and the dredge hauled, contents emptied onto the tray in the stern and the oysters picked out. Then once the cultch is shovelled back over the side onto the fishery the dredgerman is ready to row back for the next tow, and this may be repeated 30 or 40 times a day. Many years ago these punts had a small lug and mizzen in to get to the fishery, but now there are none of these left, and outboard motors are the order of the day. In light weather more and more sailboats are towing punts behind

Wink boat dredging with 'Twilight' in the distance.

them and if there is insufficient wind to sail, the sailboat is anchored on the Fishery and the punt is put to work.

Nowadays only a handful of boats work this way and are outnumbered by the sailboats which is unusual in the history of the Fishery. Usually the winkboats outnumber the sailboats by 3 or 4 to one and in 1924-25

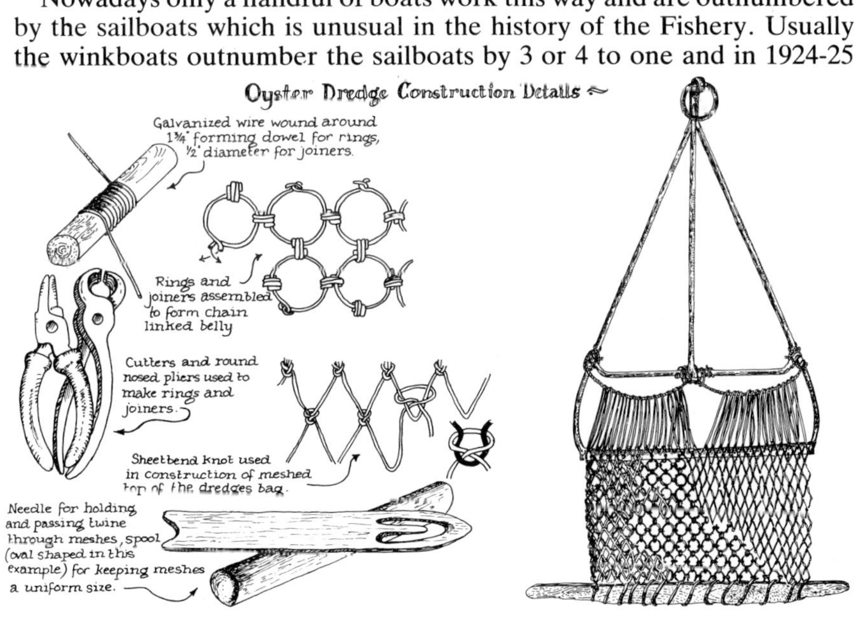

season there were 187 working as opposed to 50 sailboats. This was the greatest number recorded though up until this date they regularly numbered over 100. Generally speaking the men who operate the winkboats, also known as 'haul-tow' boats, have come from the creeks in the higher reaches of the Fal and work in the calm waters around them. The main area is Coombe and the men are known as 'Coombe Gaffers'. They usually stay up the higher reaches and work down as far as Turnaware Bar, though in fine weather they do come onto the main fishery worked by the sailboats. When there is little or no wind, many of the sailboats are anchored on the fishery and the men take to punts to try and earn a day's pay. The day begins for the dredgerman at about 7.30 am when he has to get up and start his preparations. Half an hour later he will be down at the waterfront assessing the weather for the day to come, and wondering how accurate the previous nights weather forecast will prove to be. Once he has rowed aboard he may start his engine if he has one, for it is permitted to motor out to the

'Result' and 'Evelyn' prepare for work.

fishery as long as the engine is off and sails hoisted for the first drift. If he is engineless as some still are, then he must hoist his sails and start the journey in a more sedate and less noisy fashion. Once up there the bed to fish on is selected after some thought and the boat is set up for the day's work. The dredge warps are uncoiled and the dredges thrown over the side. A dredge is a triangular frame made of ½ dia. round

An oyster dredge seen here being emptied on 'Result'.

steel bar and when stood on end the top corner stands about waist height, and has an iron ring to which the warp is fastened. A warp is usually about 15 fathoms of ¾″ rope. The base of the dredge is flat to form a knife that digs just under the surface of the cultch as it is towed behind the boat. The blade has holes drilled, through which the dredge rings are attached, and the wire dredge belly is fastened to these rings. The spacing of the ring is uniform on most dredges and used as in gauge of dredge size, thus an eighteen ring dredge would be used in a punt as it can only tow one dredge and so has as big a one as possible, whereas a sailboat towing 4 dredges may only work 15 ringers. The iron belly is towed over the ground behind the blade the net is tied over the top of the belly and the oysters scooped up by the blade are collected in the bag thus formed. The wire belly is necessary otherwise with the constant dredging and weight of oysters in the bag, the bag would chafe through very fast where it is made of twine. A wooden 'chopstick' is attached to the after edge to ensure it maintains its shape. Once thrown over the weather side and the correct amount of line paid out – a skill only acquired by experience – the warp is attached to the boat by a 'stopper'. This is a light bit of twine 6′ long which is used to operate the weak link system. If the dredge should come fast on a 'hitch' on the fishery, the thin line will part with less damage to the boat as a dredge warp would do if it were tied to a cleat, and the warp is thrown over the side. A

Here the stoppers can clearly be seen, providing a weak link in the event of getting a hitch.

buoy is permanently attached to the end of the warp in case of such an event, and so the boat merely has to sail back to the dredge and pull it up. The sails are now set up and once the boat is going well, the first dredge of the day is hauled aboard and its contents emptied up on the

The contents of a dredge, mainly dead oyster shells and a few live ones.

cultch tray that is resting on the side deck. Once the oysters are sorted and predators removed the tray is designed to tip up and empty its contents back into the river. Any oysters to be cleaned can now be dealt with by the culling knife called a 'cultack', a piece of iron flattened by a blacksmith until roughly knife shaped and which have an iron age look

On board 'Shamrock' John Old empties his cultch back on to the beds. This is an important part of process as it ensures a layer of cultch (dead shells) cover all the fishery providing something for the spat to settle on.

to them. Once the dredge is sorted the next is hauled immediately and the process repeated until the drift is complete. Then the dredges are stowed aboard and the boat sailed back to the starting point for the next drift. Maybe five or six drifts can be done in a day and then it is time to tidy the boat, count and bag the oysters, wash down, lash down anything moveable if the weather is bad and head for home.

Frank Cock demonstrates the way to test an oyster to see if it will hang, and therefore can be considered a legal oyster.

'Evelyn' and 'Beatrice' drift across East Bank.

During the months at the beginning and end of the season it must be as near as perfect a job as anyone interested in sailing boats could wish for. The weather is mild and often sunny, winds not usually too strong or too light. These are the days that we all tend to remember, but of course Mother Nature will never let us have too easy a time of things, and between the good bits there are plenty of bad ones, enough to deter even the hardiest of romantics. It is in the end just a job, a means of livelihood, and whereas it is far better to be able to have satisfaction and pleasure from a job, the first attraction must be the means of earning a living. No person in their right mind would spend a winter being lashed by gales and rain, suffering the cold months of December, January and February just for aesthetics, though many are prepared to do it for money, and few of those are not rewarded occasionally by thoughts of the beauty of the life-style.

This time Johnny Walls has a chat with the Laity brothers, Paul and 'Hammer', as 'Beatrice' and 'George Glasson' pass each other during a drift.

CHAPTER 8

Dredgemen Past and Present

As in most spheres of physical occupation, oyster dredging has produced a tough breed of men, their lives moulded by the vagaries of wind and tide and the intricacies of working a boat under sail. And, inevitably, the life-style has produced its fair share of characters and incidents that will long be remembered around the estuary of the Fal as long as there are names like Ferris, Vinnicombe, Laity, West and Gunn out on the water.

Although the dredging season is a winter job only, it must be realised that most of the men in it spent their entire lives afloat, turning their hands to a variety of other activities during the summer months, mackerelling, trawling, long-lining and even scalloping, often adapting their oyster boats for the purpose. Like all men who work hard in a frequently hostile environment, when they come ashore the fishermen relax with equal vigour and it is hardly surprising that many of the more amusing stories they recall are directly related to the consumption of alcohol, sometimes in excessive quantity! What kind of shortsighted person would ever consider starting a working boat race while the pub was still open? But, certainly, that was the case one year at Porthallow regatta where the frustrated starting man, in a final desperate attempt to prise the working boat crews out of the 'Five Pilchards' where they were firmly ensconced, let loose his twelve bore right outside the open window with a cry of 'That's your start, Working Boats!' A drastic move, perhaps, but it did have the desired effect and still stunned by the shock the crews evacuated the bar like a herd of startled cows in a running start to their race. And when the race was over, there is always a frustrating time vacuum before the next opening time, usually filled at most regattas by a break for tea. Porthallow was no exception, but it soon became common knowledge that members of the band were entitled to free teas and it was not long before their smart hats were quickly appropriated by the crews as they came ashore. But most teas have to be paid for, so imagine the shock that awaited Arthur West on one occasion during the tea break at Portscatho regatta when he became separated from the rest of his crew aboard 'Victory', who got into the cafe first, and informed the proprietor that "Mr. West would settle up when he arrived", piling their plates in the meanwhile with cakes and sandwiches. The rest of the working boat crews soon caught on to this bonanza, and when Arthur finally appeared he was presented with a bill for no less than forty teas! Whether his teatime generosity was later repaid in the pub is uncertain, but it is highly likely that it was albeit in a liquid form!

Still there 50 years on, a different spinnaker but the same knot on 'Victory'.

Authur 'Toby' West and the Victory and her racing successes can only be described now as legendary. Apart from her numerous cups, the Victory must also rate a special distinction as the only Working Boat to have ever graced the pages of the 'Times', the result of a classic incident

It is only due to the dedication of the crew that an old boat like 'Victory' can withstand the strain of racing, it would cost too much to have a job like this done in a boatyard. Here 'Mally' Wales takes on a major repair with the help of Robert 'Woody' Woodhouse.

that really epitomises the whole essence of Working Boat racing, which took place in Fowey between the Wars. As there was no formal Working Boat class that year Arthur raced up to Fowey along with the rest of the crack yacht fleet, and created not inconsiderable embarrassment when he showed a clean pair of heels to the rest of the fleet, winning the race easily. With many thousands of pounds invested in their specialised racing machines there was also considerable indignation among several of the yachtsmen at being seen off by 'an old fishing boat' and one owner, Mr. Holman was sufficiently moved to demand that the 'Victory' be properly rated at Lloyds of London if she was to continue racing with the yachts, announcing that he would pay for the entire cost of bringing a surveyor all the way from London himself. The man duly arrived and his survey was proceeding admirably until he came to measuring the 'Victory's' spinnaker and discovered the huge knot tied in its head. As with most working boat spinnakers it had been acquired from a yacht, and, being rather too long in the hoist, rather than have it recut, Arthur had merely whacked in a nice big overhand knot to take up the slack, the normal working boat solution to such a problem. But in the rules and regulations of the Lloyds rating system, no such eventuality was even considered, it was impossible to measure and the Surveyor had to admit defeat, returning in frustration to London unable to complete his job, leaving Mr. Holman considerably out of pocket, and Arthur still

A happy crew on 'Victory'.

out in front, knot and all! And, somehow, the 'Times' got hold of the story and if those yachtsmen had been embarrassed at being beaten to Fowey by 'the old fishing boat' imagine how they must have felt when they opened their paper that morning!

Undoubtedly, Arthur West is one of the 'characters' of the Working Boat racing scene, but in contrast, the dredging itself has produced a few, and if you ask any of the older men out on the river who they remember there is inevitably a wry grin, and something along the lines of....." and of course, there was Beecham Pill, 'Our Saviour'.....Tommy Simmons." And it is hardly surprising that they do remember him and his boat, the 'Fairy' for Tom Simmons was proof enough that the present day phenomenon of sponsorship is nothing new, as he readily accepted from Beechems the offer of a new mainsail every five years, earlier this century in return for having "Beechems Pills they're worth a guinea a box" emblazoned across it by his sponsor. But there are other things they remember Tom Simmons for, and nowadays some research psychiatrist could possibly have a field day preparing a paper on the almost human relationship Tom Simmons seemed, at times, to share with his boat. On one occasion, apparently, he missed, or rather, the '*Fairy*' missed the mooring, and after first beating the boat with the tiller, which snapped in two, (causing him to sail into his punt as he searched for a spare) he decided to further punish the boat by making it sail all the way out of the harbour and around Black Rock! He was undoubtedly a man of short temper; making his way down the beach to his punt one day in old style leather seaboots, he was seen to slip on the seaweed and rising painfully to his feet again he vent his fury on the unfortunate punt saying, "Try to drown me would you, you bugger", and by kicking it hard in the bow. Perhaps it was rotten, or perhaps Tom Simmons had forgotten the steel toecaps .... whatever, his kick smashed a hole clean through the planking. And if he seemed to make a habit of taking out his frustration by punishing his boats, there was one occasion when the 'Fairy' got her own back, as Tom went aboard early one morning and jumped down into the boat from the side deck. The floorboards were tired, to put it mildly, and Tom was a big man. He shot straight through them and the passing crew that responded to his belated cry for help found him trapped up to his waist in the bottom of the boat, "looking like he'd just ascended from below!"

When the thole pins in his punt broke, rather than go to the trouble of finding new ones he simply solved the problem by cutting notches for the oars into the gunwales. Poor old Tommy was in trouble yet again at the start of one oyster season, when after throwing over his four oyster dredges, did he suddenly realise too late that he'd forgotten to make fast the dredge warps.

And, certainly, if there was any romance to the task of earning one's living under sail, Tom Simmons did not imbibe it. The late Norman Ferris recalled the day when, as a small boy who could think of nothing else but getting afloat, sailing and boats, he chanced upon Tom Simmons busy caulking and tarring the bottom of his boat on the hard at Restronguet Pond and anxious to make contact with one of his 'heroes' he approached full of questions. With Tom Simmons' skin burning from the tar and gasping from the strong fumes it was undoubtedly an inappropriate moment for the young lad to make his approach. However, blinded by youthful enthusiasm he politely enquired, "Excuse me, Mr. Simmons, but how old is your boat?"

Tom Simmons was not exactly sympathetic in his response. Cleaning off the mixture of tar and water that was pouring out of the boat onto his hands, by wiping them on his beard he straightened up, glowered at the young, expectant face and roared, "I don't know, Boy, and what's more, I don't give a bugger!" And thus one more childhood illusion was cruelly shattered.

There were other characters that Norman Ferris was able to recall, not the least his uncle, 'Bummer' Ferris, who acquired his curious nickname from his business of sailing out to anchored ships with stores and then dropping heavy hints about a bit of canvas for his old mainsail as it was getting a bit worn, or sail twine if he did't think there was any chance of buckshee canvas. He also used to deliver stores to the Royal Naval Training ship 'Ganges' then moored in the Roads, and was in fact their butcher. He would take half-a-dozen pigs out at a time which he would first despatch outside the Pandora Inn before delivering them, no doubt a sobering experience for the regulars drinking inside! When 'Bummer' was sailing through the fleet it paid you to keep a weather eye open as he was very shortsighted and with his metal rimmed glassed and great bushy beard according to Norman, he looked like 'a bird peering through a bush!'' 'Ripple' did not have a great deal of draft so 'Bummer' concocted a rather unorthodox 'centre plate' which he lowered up and down by means of a chain attached to the transom.

Frank Cock is another man from that generation and in 1980 he was still fishing the 'Morning Star', then the oldest boat in the fleet. That she survived was undoubtedly due to Frank's efforts for when he bought the boat way back in the 1950's she was in very poor condition and he recalls one morning when he and his partner Harry Barnes were dredging soon after they bought the boat. Harry was taking a break sitting down on the floorboards eating his mid-morning 'crib' when Frank asked him the time. Without looking up Harry replied that it was about half-past ten. Intrigued, because he knew that his partner did not possess a watch, Frank asked him how he knew and Harry replied that he could see the mail van climbing Mylor hill. "But you can't see that from where you're sitting down there Harry!" protested Frank, even more intrigued.

"Oh yes I can!" retorted Harry. "I can see clean through the gaps in they planks, Frank!" The tale is further embellished by Frank's recollection that they were bailing sixty buckets for every drift they made with the dredges, and it was not long before they put her ashore for major repairs.

But if Frank and Harry were having to work hard keeping the 'Morning Star' afloat, at least she never completely sank under them as she did in the early 1900's when in the ownership of the Ferris family.

The circumstances of the story are vague but what makes it noteworthy was the unique cargo the boat had aboard at the time for she was laden to the deck with furniture! It is not certain who exactly was moving house but the object of the exercise was to move the furniture from Pill Creek to a new home at Restronguet. Alas, before the 'Morning Star' even left the moorings she sank, probably as the extra weight aboard pushed her aging topsides lower than usual and the water began to pour in through one of those seams that Harry eventually watched his mail van through.

Perhaps the most spectacular incident ever involving the 'Morning Star' took place, inevitably, at a regatta, in this case, at Penryn. Turning very tight on a mark just off the quay, the mass of spectators ashore suddenly found themselves rather more involved in the racing than they really anticipated as Frank jibbed closer to the quay than intended with the net result that the boom picked up two people and flung them into the water. Fortunately, no-one was hurt, but it was a race that they, and Frank Cock will always remember ....

One race that Kenny Corke will never forget took place far from the open water of Carrick Roads. Kenny, who was brought up aboard the 'Evelyn' by his father George, and who have both enjoyed many many successes in her decided to go to a well-known Holiday Camp one year and while he was there he got inveigled into joining in the racing on the boating pool in a fleet of Mirror sailing dinghies. And there the disaster happened as this top racing man fell foul to defeat by a seven year old boy. Kenny admits that for the first round of the race he was not really taking things very seriously but when it dawned on him that he was not catching up with the little lad in front, his natural racing instinct began to catch hold but try as he could, it was too late then to make amends.

Kenny Corke and the 'Evelyn' are of particular interest in these days of womens liberation and emancipation for she is one of the only recorded boats to have had a woman dredging aboard, when Mary Priddow spent a season working with Kenny. The only other woman to have worked in this very male domain was Sheila Rowe, who used to work an oyster punt by herself out of Mylor Creek. Sheila also became something of an institution in the racing scene as every year at Mylor Regatta she used to lay on an enormous tea for the working boat crews at her cottage opposite Tregatreath.

Many stories can be told by, and about Arthur 'Toby' West. He is a classic example of the 'living legend', known, loved and respected by all Falmouth and a great deal further afield too. He and his family are intertwined with the maritime history of Falmouth, as lifeboat men, quay punt men, oyster dredgermen and fishermen. One of three brothers, 'Toby' is the only one still racing Working Boats, and in June 1980 retired as Coxswain of the Falmouth Lifeboat, having served with

ARTHUR "TOBY" WEST

distinction for thirty three years, culminating his career as Coxswain of the most advanced and expensive Lifeboat in service in Britain, the 52ft Arun Class 'Elizabeth Ann'. With a top speed of 30 knots she makes quite a wake as she roars past, as the author of this book found to his chagrin. ... "It was a quiet winters morning and I was drifting slowly to work in the 'Florence', enjoying a rare day of winter sunshine and nursing a hangover with the help of a cup of coffee. The last person on my mind was 'Toby' West. However, he was out on engine trials in the Lifeboat, and I was very much on his mind. The reason for this was that I had been at his house the previous day, and seeing a pair of woodcock lying waiting to be plucked and drawn, I offered to do the job for him. He readily accepted my generous offer, and whilst I headed for home with them intending to return with them later in the day, preparations were made for the Sunday Roast in the West household. All would have been well had I not decided to 'have a quick pint' on the way, after which I had a few more and forgot all about woodcock. When 'Toby' spotted me across the harbour, he knew just what he had to do, and opened the throttle wide. It was as I saw the Lifeboat heading for me at 30 knots that I remembered 'Toby's' birds for the first time, and that they were still lying in my scullery. He turned the Lifeboat at about 20′ from the 'Florence', but the wake, a 6′ wall of solid water kept coming, and as she tossed in the sea conditions that had been transformed in seconds from those of a flat calm to a Force 10 gale, and the two of us aboard rushed to save our flasks and cups, we just caught a glimpse of 'Toby's' cheery smile and wave as he hurtled away into the distance. Revenge must have been as sweet as those woodcock which I made sure he got that afternoon as soon as I got back from dredging. I knew better than to forget a second time!!"

Perhaps, by way of evening up the score, I can include a story of 'Toby' many years ago in Fowey during 'Fowey Week'. The story is well known by all Working Boat men, but 'Toby' SAYS that he can't remember it. The story goes that whilst approaching Fowey in the early hours of the morning, a visiting yacht saw a small boat drifting out to sea on the ebbing tide. Upon fetching up alongside, it became apparent that the dingy was not quite empty, for lying fast asleep and snoring on the bottom boards was none other than 'Toby' West, who was then towed back into Fowey harbour. When confronted with this evidence, 'Toby's explanation that he was very tired after a hard sail up from Falmouth and fell asleep whilst rowing back to the 'Victory' after closing time at the pub. In spite of a nagging doubt in most of our minds, for 'Toby' is known to enjoy a pint or two, we will give him the benefit of the doubt and enter his version as the official one!!

Another member of the 'Victory' crew to get into difficulties, again after closing time, was the well-known 'Boyo' Carne. After a typical Working Boat night out after one Mylor Regatta, he and fellow crew member Trevor Wright set off for Falmouth by foot, and decided to take a short cut through the graveyard at Mylor. They were talking away to each other, when Trevor realised that 'Boyo' was no longer with him and he was talking to himself. He looked round and as there was no

'Boyo' Carne on 'Florence' during the Silver Oyster Race 1979.

sign of 'Boyo', he decided to retrace his steps and found him lying on his back in an open grave that was due to be filled the next day!!! It is a common sight to see Working Boat men looking like corpses the day after a Regatta night out, but this is the nearest any of them have come before daylight!

But to return to 'Toby', when in the right mood for yarning he is difficult to stop and a delight to listen to, as he tells of boats and men long gone. One amusing anecdote follows another, such as the novice starter on the Committee boat at Portscatho Regatta many years ago, who shouted at the Working Boat fleet manoeuvring for position at the start ... "Get in line you Working Boats, get in a straight line will you!" or whilst thinking of starters he is remindered of another at St Mawes, who upon firing the shotgun to start a race, discovered too late that it was loaded with a live round by mistake instead of a blank, and though he managed to miss the Working Boats, shot a hole clean through a Sunbeams mainsail.

If there is one charge that can never be levelled at 'Toby', it is surely that he is a poor sport, though this very virtue backfired on him during one race. He recalls .... "We were all racing as one class many years ago before the class was split into two classes, scratch and handicap, I was sailing 'Victory' on a reach between St. Just and St. Mawes and was passing the smaller 'Endeavour' owned by Henry Merrifield, when her staysail split in a strong gust. I felt sorry for Henry so I turned 'Victory' round and went back to offer him a spare staysail I had on board. He thanked me very much but waved me on saying that he could manage alright without one. "But," recalls 'Toby' with a laugh, "the joke was that I never caught up with him again, and he beat me home over the line just under jib and mainsail!"

Such is the nature of the man, and it seemed only fitting that when the Queen visited Falmouth in 1977, it was 'Toby' who was accorded the honour of taking 'one of her boys' sailing whilst she went about her Official business. So, after first being presented to the Queen in his capacity as R.N.L.I. Coxwain on Prince of Wales Pier, he was then whisked aboard the waiting 'Victory', where her crew, plus Prince Edward, were waiting for him to take them sailing. An impromptu race was organised with the 'Winnie' who just happened to be in the vicinity, and a challenge was thrown ... a pound to the first boat from Black Rock to Prince of Wales Pier. 'Toby' handed the helm to the young Prince, then only 14, and in spite of being buzzed by a catamaran ... which turned out to contain his brother, Prince Andrew ... just managed to beat 'Winnie' over the line. Some weeks later the stake he had won was sent to Prince Edward by 'Toby', and soon a reply came back from the Prince saying how much he had enjoyed his day out and his first trip

'Toby' West (in the white hat) discussing racing tactics with the young Prince Edward on board 'Victory' in 1977.

on a Falmouth Working Boat, and he enclosed 50p for 'Toby' as he felt that he was entitled to half the prize money!

Another great character, still very much with us, is George Vinnicombe. Known popularly as 'The Old Man' or 'Chan', (after a certain Chan Canasta who was also a magician, though not in boats as George is). He has become, like 'Toby', a legend in his own lifetime though his reputation has been built on his great feats as a fisherman. Every type of fishing he attempts he excels at, and he is known throughout the West Country as a Master Fisherman, whether he is fishing for mackerel, trawling, wreck-fishing, or oyster dredging. A brief glance through the pages of this book will provide evidence of the number of Vinnicombes that have been oyster dredgermen and fishermen, owning such boats as 'Six Brothers', 'Boy Willie', 'Lily of the Valley', 'Phantom', 'Shadow', 'Ziguener', 'Boy Phil', 'Honour', 'Softwings', 'Silent', 'Vera', 'Shag', 'Queen of the Bay', 'Myrtle', and probably many more besides. For many years George raced the 'Six Brothers', with a great deal of success, but when she was tragically lost in 1968, a lot of his enthusiasm for racing went, though he still dredges every winter in the new 'Six Brothers' he had built at Plymouth. George has the reputation of being a hard man to sail under, and this is borne out by the following story. He relates ...

"We were racing one afternoon in a strong wind", he recalls, "and one of the crew kept saying we should have the big jib on her. I could see the wind was increasing, and so I said that I didn't think it was a good idea.

'Six Brothers' carries her big jib, few boats carry sails of this size nowadays as they are considered less efficient than high cut jibs.

But he wouldn't take no for an answer so I said, 'If you want to put it up go ahead, but it'll stay up until the end of the race', and so up it went." Too late now the crew realised his error of judgement, and suggested to George that they take it back down again. "But," said George, "I told him that I would keep it up to the finish if he set it, and I did, though I nearly filled her up doing it. Still," he smiled, "he was never quite so keen to set the big jib after that!"

If space allowed, a book could be written on such men as these, for there are many other local families abounding with men of great character, families such as the Ferris's, Laity's, Morrisson's, Gunn's and many more.

The colourful character of the men is reflected in the colourful nicknames of some of the boats, some of which have explanations but some of which have been lost in the passage of time. Names like 'Galloping Donkey', the nickname for the Flushing boat 'Edith' owned by George Monk. She was a Mevagissey built boat, 22′ and completely black varnished except for green bulwarks, leaked like a sieve and so had to carry a 4 gallon bucket all the time, and earned her nickname by being so slow. Another Flushing boat, owned by the Johns family, was known as 'Fat Johnny', though her real name is no longer known. This nickname was derived from the fact that she used to collect tallow and cordage from the square riggers in the bay. A more puzzling nickname was that given to the 'Susan Jane'. She was one of the last boats to be trenailed, that is fastened with wooden nails, and both she and her owner were known by the nickname 'Rigsabunky', the spelling and meaning of which can only be guessed at! Her owner, one Herbie Richards, apparently used to whitewash her topsides until the early 1930's when she fell apart at St. Just on the beach.

A strange boat, if not a strange name, was the 'Pioneer', registered for oyster dredging in 1869. She is entered in the Port Registry as being 23¼ tons, having a 50′ keel, and Spritsail rigged. she was worked by a crew of three. This class of vessel, however, does not appear to have caught on!

Another great waterfront character, though more famous as a Quay Punt man, was Norman 'Ferret' Morrisson. He did however spend a little time dredging in the 'Serene', but more affectionately known as 'Black Bess,' 'Ferret' has his own solution to the problem of getting the dredges over the bulwarks ... he simply cut the bulwarks off with a saw upon returning from fishing one day, and that was the end of that particular problem!

However, great though these characters are, or were, there are plenty more to take their places, and are practising now! It is the nature of the job that cultivates the nature of the man, and as the way of life survives, so will these colourful personalities and their like.

CHAPTER 9

Conclusions and Future

The past 5 years have been very poor for the Truro Oyster Fishery, the reason being that in 1982 the Fishery became infected by a disease called Bonamia. It resulted in a Ministry ban on the movement of all oysters for the purpose of relaying outside our own area, 14a, which covers oysters caught between Dodman Point and the Lizard Light. This effectively closed the Fishery as the merchants outside this area, who had traditionally bought the Fal oysters, were now no longer able to do so.

Neither did the local merchants wish to buy the infected oysters as Bonamia is a disease that affects relaid oysters very badly and the risk of losing the entire stock was too great for them to take. In December 1981 the decision was made to close the Fishery and it was not reopened until 1984, but the reopening of the Fishery did not produce a fleet of dredging boats as the ban on relaying outside our own area was still in force, and local merchants were not prepared to risk buying in for relaying as stocks were still very low. Some of the dredgermen have found outlets for their catches by selling direct to the restaurants, this has meant setting up purification tanks and catching only the largest oysters on the Fishery, but the fact that far less are caught is compensated by the higher price that is realised. Whilst Bonamia makes the relaying of oysters impossible it doesn't affect the restaurant and hotel market as they are perfectly safe to eat even if they are carrying the disease, and neither the health of the person eating them nor the flavour of the oyster are in any way affected. Naturally the fishermen are wary of returning to the Fishery yet as many lost their entire seasons catch which they had relaid on their private layings, still a few boats returned to the Fishery for the 1986-87 season (7 sailboats and a few punts) as the disease showed signs of decline. Even so, any optimism at this stage may be misplaced as the situation is still precarious and a further onslaught of Bonamia could spell the end of the Fishery.

The Fishery has certainly had its share of trials and tribulations in recent years as in 1976 there began a rapid increase in the fishing effort in the South West approaches to the English Channel. By 1980, 15 huge foreign factory freezer trawlers (more commonly known as 'Klondykers') were operating within the Carrick Roads processing mackerel. In addition to these there were many more ships anchored further out in Falmouth Bay, some a great deal larger than the ones in Carrick Roads, but all with the same purpose in mind – namely to plunder the last of the large mackerel stocks in the world. This fleet of 'Klondykers' was comprised of vessels from 200′ to 600′ from

A Scottish purse seiner landing to a Klondyker on the edge of the fishery.

many countries such as E. Germany, Poland, USSR, Egypt, Spain, Italy, France and even a few from the UK! This international fleet was supplied with fish caught by a fleet of about 100 'catcher' vessels varying in length from 80′ to 170′, most of which were from Scotland. As a result of these activities the Carrick Roads became badly polluted in the area of the oyster beds in two different ways. Firstly from oil, which was sometimes the result of accidental spillage, but all too often as the result of boats pumping their bilges out into the harbour instead of waiting until they are back at sea. If oil settles on the oysters as they are exposed at low tide it is fatal for them, and due to the difficulty in proving which of the hundreds of boats was responsible, the task of claiming compensation would be an impossible one. The oysters on the Fishery are safe from this type of pollution as long as they never uncover at low tide and the oil does not come in contact with them, though oysters relaid onto layings on the foreshore are vulnerable as they are exposed at low tide.

The second type of pollution was from mackerel that are dumped on the sea bed if a 'catcher' boat had in excess of what it could sell, or if the mackerel were not of the size required by the freezer trawlers. This type of pollution poses as great a threat, if not greater, than by oil pollution as the oxygen levels in the water drop whilst the fish decompose, resulting in the death of many oysters and other shellfish from oxygen starvation.

The critical level is about 85% of the normal level, once it drops below this oysters start to die and it has been below this level on several occasions. The dumping of unwanted mackerel has the dual effect of reducing the oxygen level as they decompose, and making large areas of the Fishery unworkable as the dredges come up full of stinking, decaying mackerel instead of oysters. It may be true that the mackerel fishery has provided the country with millions of pounds worth of valuable exports, but there seem to be few benefits for the Cornish people who live on the River Fal and have had to put up with the inconvenience of noise, smell and pollution. In fact many dredgermen are mackerel fishermen who were forced to sell their boats as a result of the overfishing by the 'invading' mackerel fleet, and once again found their livelihood in jeopardy up on the oyster Fishery.

The threat has receded for the time being as the mackerel stocks were so badly depleted by the overfishing during the late 1970's and early 1980's, that by 1982 they had all dispersed to find richer pickings elsewhere, and the Government 'shut the stable door after the horse had bolted' by putting a ban on mackerel fishing by any method other than handlining, but only after the once profitable handline fishery had been wiped out. It became evident to all the dredgermen that their Fishery was not considered to be of any consequence in the eyes of the bureaucrats and the risk of killing off the last natural oyster fishery in Britain would not be allowed to hinder their quest for the 'mighty dollar'. Doubtless the vast investment in the UK mackerel fleet and the powerful lobby their owners were able to muster easily overcame the protests of the thousands of men involved in the Cornish mackerel fishery, what chance then for a few dozen dredgermen to make their protest heard. There is constant pressure from the owners of the 'Klondykers' and the Scottish purse seiners to regain access to the South West mackerel stocks which are at present at a very low ebb with fishing only open to small fishing boats using handlines. If the stocks start to recover then the constant pressure being applied will certainly intensify and doubtless the interests of the local fishermen will be ignored yet again and the Truro Oyster Fishery will once again be under seige from pollution.

Maybe if they had paid more attention to the insignificant little oyster fishery with it's quaint, old fashioned sailing craft, they might have learnt a valuable lesson about conservation and perhaps the South West mackerel stocks might not have been decimated for the sake of only 4 or 5 short years of greed. It may be a fluke of circumstance that the Truro Oyster Fishery has stumbled upon the secret of conservation through inefficiency, but nevertheless the results are there for all to see. Though it may be too late for many of our traditional fisheries (such as

Should we still call in progress when one boat like the one above with a crew of only 15, catches as much mackerel in one day as the fleet of handliners below can catch with 40 men in ten weeks?

herring and pilchard in Cornwall) which have been fished to the point of extinction, there are still plenty of fish left in the sea and it's not too late to take a less aggressive approach if we can learn to contain our greed. The difference between efficiency and overfishing is both subtle and lethal, so no matter what examples we see set by other fisheries in the name of 'efficiency', we must hope that the Truro Oyster Fishery will be allowed to remain both inefficient and prosperous.

An even more deadly threat is posed to the survival of the Truro Oyster Fishery by the proposed building of the Falmouth Container Port. It was possible that this development would sound the death knell for the Fishery, and in spite of requests by the Dredgermens Association for information we were dismissed as small fry. If we had little chance of a hearing from the Ministry over the mackerel fleet, then we had absolutely none with the powerful interests involved in the Container Port development.

The first reason for concern is that during the massive land reclamation scheme large quantities of silt would be deposited upriver onto the Fishery. Oysters are very susceptible to silt as they are filter feeders, and once the gills through which they feed become clogged up (as they do with sand) they can no longer extract the food they need from the seawater and so die.

The second worry was that as the proposed Container Port was to extend out from the Eastern Arm of Falmouth Docks in an easterly direction, thereby reducing the harbour to about two thirds of its present width, the resulting effect upon the speed and direction of the tide and currents could only be guessed at. A very delicate balance of current, water temperature and food supply are vital to an oyster fishery and a small change in any one of these factors could have dire consequences. It is possible that the inevitable changes in tidal speed and current directions would have little effect on the Fishery – who knows, they may even have proved to be advantageous – but as the Fishery has survived so many years with the present balance of conditions, it should be assumed that they are perfect and any change in them is more likely to be detrimental than beneficial. It was too much to expect that the new developers would provide the dredgermen with some assurance that their interests would at least be considered, and that the dredgermen would be able to discuss their anxieties over the affects on their Fishery with them. It looked for a time that yet another small industry was about to fall victim to 'progress' without even so much as the courtesy of consultation or consideration, and so not only the dredgermen but also the majority of people in Falmouth are greatly relieved that this project appears to be dead and buried at long last after hanging over our heads for the last 20 years. It may well be that the old and new can co-exist if some attempt is made to reconcile the needs of the two industries, but as in this case it seems that the new are rarely concerned with making any such gestures, their own needs being paramount.

The racing fleet has its share of storms ahead too, the first being the decline of the Fishery, for should the Fishery succumb to the difficulties it faces at present, then the racing fleet will provide little interest save as another fleet of gaffers 'that used to work'. It is vital that the Fishery be preserved if the racing fleet is to maintain its unique character and continue its development, the name Falmouth Working Boat won't have quite the same ring to it if the boats don't actually work!

Another possible danger is from commercial exploitation, with the cost of gear and sails soaring and the standard of racing gear rising every year, the time is fast approaching when it's going to be too expensive for working men to fit out their own boats for racing. It could even be argued that this point was passed some time ago! Many of the racing boats now have to operate as a syndicate in order to help the owner meet the cost of sails and gear, so surely the next stage will be sponsorship. If this were to happen a great deal of thought will have to be given to the implications, for once we go down that road it may prove very difficult to turn back if it is not working out the way we envisaged. Once one boat accepts sponsorship it will have such an advantage over the others that they will have to follow suit in order to remain competitive – and what happens to the boats that are unable to get sponsorship! Maybe this is a good thing, perhaps not, but what is important is that the problem be recognised and discussed. Should it be felt that sponsorship is the right course, it should be given to the Association and not individual boats in order that it can be shared equally through the fleet. This would surely be better than a 'free for all' with individual boats seeking sponsorship. One point in favour of sponsorship is that it would help to preserve the character of the class in that it would allow the less wealthy owners, such as the dredgermen to compete and keep them interested in joining in the racing, as very few are at present.

The problem of sponsorship is still a problem of the future, but one that is very much with us is that of the fibreglass boats and the role of the Association. This is a very complex area which is being ignored because it is so difficult, but unfortunately it will not go away and we will only bring far greater problems upon ourselves if we do not tackle it as soon as possible.

The introduction of fibreglass was welcomed by the Oyster Dredgermen as the rising cost of wooden boats was making replacement of old wooden boats uneconomic in terms of returns from the oyster fishery. The new fibreglass hulls proved to be every bit as efficient as their wooden counterparts, and were probably about ⅓ of the cost. This first revolution within the Fishery was, and still is a success, but has given the racing fleet a dilemma that it still has not resolved after 20 years. The problem can be broken down into two parts, let us consider them one at a time.

1. The difference between the racing abilities of wood and fibreglass Working Boats.

There can be no doubt that if two identical boats race each other but one is made of wood and the other of fibreglass, the fibreglass has the advantage. The reason for this is simply that the fibreglass boats are able to get their ballast lower in the boat, thus lowering the centre of gravity and enabling them to sail more upright, and therefore faster. They are able to get the ballast lower for two reasons. Firstly, as the weight of a fibreglass hull is considerably less than a wooden hull, it follows that in order to make up the same overall weight a lot more weight can be put on the bottom of the hull. Secondly, the latest technique is not to fix an iron ballast keel in the traditional manner, but to build only the shape of the ballast keel as part of the hull moulding and to then fill this from the inside with iron punchings or lead and then fibre glass over the top. This makes sound sense from a shipwrighting and safety point of view, but unfortunately from the racing point of view this gives the fibreglass boats even more of an advantage. A wooden boat cannot have punchings moulded into the hull in this fashion, and so has to be content with laying iron ballast pegs on top of the floors and frames, which naturally heap up much higher than do the punchings. These two factors combine to give an advantage to the fibre glass boats before the race starts. However, as the feeling of most of the skippers is that we are better off racing as one class rather than two, or introducing a handicap system for the Scratch Boats again, it is a problem brought on by progress that we will have to live with.

2. Where to draw the line as far as racing boats are concerned.

Ever since the racing first started, there have been boats racing in the class that have not been Working Boats in the strictest sense of the term as they weren't raced by dredgermen, but they have always been in the minority and were normally boats that had worked in previous years now purchased by new owners who wanted to use them as yachts as opposed to fishing boats. These boats have always been readily accepted in the class, and in fact have often done the class a service, as when they are bought by such an owner he normally renovated the boat to a standard beyond the pocket of most fishermen, and when they tired of their new toy some lucky dredgerman could buy a boat in first class condition at a good price! They were always in the minority and so realised that they were 'the guests of the Working Boat Class', they normally appreciated the 'spirit of the class' making no attempt to alter things. This tradition has continued right up to present times with such boats as 'Winnie' being raced by local dentist Joe Keast, 'Mildred' by local solicitor David Stonehouse, and several others besides. All these boats and their crews have proved an asset to the class, and it is only right that the old boats should be kept alive in this way even if they are no longer dredging. The situation that we now face is that there is an influx of the newer fibreglass boats, boats that have not been built for dredging, by men that have no connection in any way with the Working Boats, built purely for the purpose of racing. Strictly speaking they are not Falmouth Working Boats, but yachts built on the same lines and as such should not have been allowed to race in the class, but dredgermen are of an easy going nature and let them join the racing. The trouble is that having let some in, the precedent had been set, and because of the number of potential yachts that could demand entry into the class, it was decided to form an Association with the stated aim 'to preserve and encourage the unique tradition of the Falmouth Working Boats' in order to try and keep the class pure. To date this has not proved to be the success that was hoped for, partly because the rules were not too well defined and it is still possible for a boat that was not built to dredge (and has no intention of

The death of a lovely old boat, 'Morning Star', seen here at Tregatreath Boatyard, is damaged beyond repair.

doing so) to join the class, and partly because of the dictatorial nature of the Association, which has prevented several dredgermen from racing their boats as they refused to join the Association. This was doubly annoying as whilst these boats were laid up watching the racing, their places were taken by boats that were built purely as yachts for racing, and sailed by men that will never dredge them.

There are several problems created by allowing these 'yachts' into the class, the first being that if we let them all in they will become a majority and in a position to dictate to the dredgermen, whose class it is, and effectively take over the class altogether. The point is that it is the men that make up the character of the Working Boat fleet as opposed to the boats, and if the men are 'diluted' in this way so is the character of the class, which once lost can never be regained.

Secondly, the racing is of a unique nature, because of the high standard of boathandling amongst the working boat skippers who spend their winters sailing for their living, combined with a disregard for all but the most basic rules. Working Boat racing has traditionally been done by common sense and the 'seat of the pants' sailing and the only rules normally used were the Port and Starboard Rule and giving way on a mark if the other boat had an overlap. All the rest of the rights of way would be decided by a combination of sometimes common sense, sometimes courtesy, and sometimes bluff. It made for exciting and original racing, but this is all being lost now as more and more 'rule orientated' yachtsmen join the ranks of the Working Boat fleet, protests once unheard of amongst Working Boat men are now becoming the order of the day and a situation is fast arriving where a race may be won by a superior knowledge of the Rule Book rather than sailing ability. A keen yachtsmen who has spent all day in the office will look forward to getting out on the boat as soon as he leaves work in order to 'tune' his boat up for the next race, whereas the dredgermen who is probably a fisherman in the summer has spent a long day at sea and the last thing he wants to do is go on board another boat. The part-time sailor will have the enthusiasm to spend his spare time looking for hollow spars, lighter rigging etc., and having found these things will have the time and inclination to fit them to his boat to aid his search for those few extra seconds of boat performance. Not so the fisherman, but he is starting to find himself at an unfair disadvantage unless he follows suit.

The wealthier yachtsmen can afford to take chances carrying sail in heavy weather, as a repair to him whilst it may be inconvenient, poses no financial problem as it would to the dredgerman and the boat will be ready for the next race. Similarly a spar can be bought if one is carried away by excessive sail carrying, whereas the fisherman would have to make his own. These factors are of course borne in mind by the fisherman when deciding whether or not to reef in a breeze, giving yet another edge to the 'yachtsman' as the fisherman's pocket usually dictates that he errs on the side of caution.

Furthermore, during the winter months when the oysters are in season and the dredgermen are at work, the 'yachts' are laid up and polished in readiness for the next seasons racing, not for them the worry of trying hastily to repair the damage done to the 'racing finish' by the heavy dredges in time for the rapidly approaching racing season whilst working long hours fishing for a living.

Add to these advantages the fact that the 'yachtsman' is sailing one of the superior fibreglass hulls, then it becomes apparent that this is a situation that must eventually explode. Is it acceptable to open this unique class of boats to any yachtsman who fancies 'a go at the working boat boys'? Of course not, but this is the situation at present and can only result in more of the genuine working boat men losing interest in their own class and leaving it to people who seem to feel the need to become part of the Working Boat fleet, but blind to the fact that by becoming part in the way that they do, they are destroying the very nature of the class, the thing that attracted them in the first place.

So the Association will crumble, for what use is the Association if it does not command the following of the entire Working Boat fleet? There is no doubt a need for an Association in order to co-ordinate the racing which has now grown to a point where we have 3 or 4 times the number of races that we used to have, and to help with the raising of funds to help meet the high cost of insuring these boats, but if gradually the fishermen pull out all that will be left is a hollow shell run by imposters. Genuine Working Boat racing will never die out, for the men will always want to race and will merely start up again somewhere else once enough of the 'proper Working Boats' have left the Association, but it seems a terrible waste to let this happen as so much effort has been made to get things as far as we have to date.

Is there a solution to the problem? There are of course several, but most are not really the ones that we want. For instance we could split the class between wood and fibreglass boats but this would create more problems than it would solve. How can a man who dredges his fibreglass boat in the winter be told, in all fairness, that he may not race with the rest of the dredgerman in wooden boats as he has done in the past, and that he has to race in a fibreglass class!! It is hardly his fault that economics dictated that he should have to have a fibreglass boat instead of a wooden one.

We could put a time penalty on the faster fibreglass boats, but as we have only in recent years split the class into Handicap and Scratch boats, it would be a step back that no-one will want to take as the two classes have proved to be such a success.

We could make a rule that only boats that are dredging may race, but rules are made to be broken and if we are not careful we will only succeed in getting more boats on the Fishery simply in order that they may be eligible to race, thus putting extra pressure on the Fishery and dredgermen trying to make a living, or a boat might just do one days dredging and then try to claim the right to race. Another important consideration is that several of the genuine dredgermen not dredging at the time could find themselves prohibited from racing by such a rule, and anyway, as we have long had a tradition of non dredging boats in the fleet, we should be able to continue this tradition at our discretion.

So is there an acceptable solution? If we analyse the character of the Falmouth Working Boat class it is clear that the common factor is that all the men or boats are in some way connected with the Fishery, either the boat has worked on the Fishery or the man in the boat has a strong connection with the fleet. With very few exceptions the Working Boats are crewed by local working men of the area, and so all the men are of the same background, attitudes and income. It is this that binds all the crews together ashore and afloat, and the sailing in the summer is just another part of the social life of the waterfront community. This is what we should try to preserve and if we can succeed in that we will have good chance of preserving the unique atmosphere of Working Boat racing.

Can this be done? It could possibly succeed if we were to close the class to any new entries unless they are bona-fide dredgermen, only allow others in at the discretion of a committee of Working Boat skippers, and throw away the Working Boat Association rules that are tying us down. In the past it would have been ridiculous for anyone to ask whether they could race with the class as you knew whether you were a Working Boat man or not, and if you weren't you didn't join in unless invited. Now with the advent of the Association rules which define exactly what constitutes a Falmouth Working Boat (no mention is made of the person in it!), we have reached a state of affairs where these boats which would previously never have dreamt of racing with the Working Boats are just turning up on the line at the start of the race simply because they own a boat that fits the specification laid out by the Association rules. Not only do some of these boats not contribute in any way to the social side of Working Boat racing at the end of the day, but in a few instances their lack of sailing ability has put other boats at risk. If statistics were kept they would definitely show an increase in collisions and other 'incidents' as more of due to one or two owners of dubious ability who have been allowed to race when a polite 'no' would have served the Falmouth Working Boats reputation better. The damage that has been done so far is not beyond repair though as long as we stop the rot now. If we prevent any more 'yachts' (wood or fibreglass) joining the class we may be able to preserve the true Working Boat racing instead of allowing it to become just another class of 'Old Gaffers'. There are many other classes for 'Old Gaffers', but there is only one Falmouth Working Boat fleet.

Complacency must not be allowed to creep in as this is certainly a time of change for both the racing and Fishery. Changes have a habit of sneaking in unnoticed, and it is often too late to correct the damage by the time it is spotted. After all it was only a few short years from the time the first Scottish purse seiner arrived, that we had a fleet of well in excess of a hundred boats, 20 of which were anchored within 1 mile of the Fishery, and spewing their pollution on anyone unwise enough to get too close. Unfortunately the Fishery cannot move and so has to fight for survival from where it is.

Hopefully the oyster boats of the River Fal will become better known over the next few years and will begin to get the protection they should be entitled to. If they become more of a tourist attraction they might cease to be the 'forgotten fleet' and perhaps will no longer be misrepresented in the press as 'fine old gaff rigged oyster boats that used to work under sail, and are now raced as yachts'. Let us hope that this was merely a printing error and not a prophecy.

APPENDIX

Name	Present Whereabouts	Comments
Ailsie–FH 289	NLF 1924	Aux. Gaff Ketch. Oyster Carrier
Anna		
Alfred – TO 205	Condemned 1914	
Alarm – TO 97	Reg. cancelled 1873	
Ann (ex Georgina) – TO 25		
Agusta		
Augusta		
Agenora		Dandy
Alf Smithers	Falmouth	
Agnes	Falmouth	
Ada	Falmouth	Ex Hayle Pilot boat
Boy Phil (ex Jubilee) – TO 33	Falmouth	
Beatrice – FH 53, FH 132	Falmouth	
Boy Willie – FH 326, FH 83	Left Port	Ketch
Boy Ben		
Boy Peter		Clinker built Yawl
Brothers		
Blanche – TO 38	NLF 1929	
Boy Jim (ex Ellen)	Broken up 1931 in Penryn	Dandy
Blue Eyed Maid		
Boy Warry – TO 19		Yawl dismasted 1930 and used as store ship
Blossom (ex Kindly Light)	Falmouth	
Boy Willie – FH 326	Falmouth	
Cock O'The Walk		
Chrystal Spring	Left port	
Charm – FH 124		Sold as Yacht 1945
Crest of the Wave		
Caroline		Dandy, Oyster Carrier
Cheyney Rock – TO 36	Condemned 1878	
Chance – TO 34		Sold as Yacht 1927. Had centreplate
Carrie	Falmouth	
Catherine – TO 87		
Deerfoot (ex Xerifa) – TO 40, TO 242, TO20	NLF 1920	
Dawn – TO 114, TO 245	Broken up 1892	Decked Yawl

Name	Present Whereabouts	Comments
Dorothy Louise – FH 39	Falmouth	
Dolphin – TO 64	Broken up 1883	
Dairy Maid – FH 47	Sold to foreign owners 1926	Aux Dandy
Deacher – TO 29	Sold as Yacht 1952	Smack Rig. (Smack rig meant a bigger mainsail)
Echo – TO 17		
Endeavour – TO 4	Falmouth	
Emma		Dandy. Sold to Scilly Isles as a flower boat
Ellen – TO 228 FH22	N.L.F. 1928	
Evelyn – TO 18, FH 13 FH175	Falmouth	
Elizabeth Mary – PH 107	Dartmouth	Built as Folperro Gaffer
Ellen	Reg. cancelled 1928	
Eaglet		Yawl
Edith Annie – TO 238 TO 28, FH 250	N.L.F. 1928	
Elizabeth – TO 47	Broken up 1873	
Express		Dandy
Eliza – TO 49		Went to Germany 1935. Wm Ferris put a new bow in her.
Etta – TO 90	Reg. cancelled 1873	
Ergo – TO 98, TO 2	Condemned 1913	Entry in fishing register say she was “condemned, worn out”
Edith – FH 158.		
Emma Louise	Falmouth	
Florence – TO 9	Falmouth	
Five Sisters – TO 18	Falmouth	
Fear Not – TO 118	Sold as Yacht 1947	Dandy. Fell over and broke up after a leg collapsed.
Flora		Dandy
Fal		
Fairy	Broken up 1917	
Forget Me Not		Lengthened and given a new bow by Wm Ferris.
Fairy Queen – TO 62		Dandy
Five Brothers – TO 12		
Flying Cloud	Falmouth	

Name	Present Whereabouts	Comments
Gazelle	Falmouth	
Glory		
George Glasson FH 173	Falmouth	
Gypsy FH 101	Falmouth	
George & Mary		
Girl Sarah	Falmouth	
Honour	Falmouth	Built as Newquay Pleasure Boat
Happy – TO 26	Portland	Left Falmouth in 1973
Heartsease FH 150	Sold to Polruan 1917	
Harriet – TO 240, TO 42	Broken up 1968	
Helen Mary	Falmouth	
Hope		
Ivy	Falmouth	
Isadora		
Ida – TO 14	Broken up 1950	Counter stern
Irene	Falmouth	
John Wesley		
Joseph Eliza		
Joseph Collins		
Julie	Falmouth	
John TO 66		
Jason – TO 13	Condemned 1912	Entry in Fishing Register says "Worn out"
Josephine – TO 44	Reg. cancelled 1878	
Kestrel – TO 16, FH 11	Condemned 1905	
Kathleen – FH 31	N.L.F. 1928	
Kate – TO 62, TO 239	N.L.F. 1886	Entered as "Broken Up" in Fish. Reg. 1877!
Katherina – TO 15, FH 254 FH 204	Falmouth	
Lilian – FH 43	Falmouth	
Lucy – TO 84	Reg. cancelled 1872	
Lucy – TO 45	Reg. cancelled 1878	Yawl
Lily of the North – FH 199	Broken up 1921	Sank in 1917. Yawl
Lucky Boy	Broken up 1986	
Liberty		

Name	Present Whereabouts	Comments
Little Ellen		Dandy
Lily	Left port	Counter stern
Leila – FH 360	Falmouth	
Lorna Doone – TO 35, FH 202		Yawl
Lottie – FH 11	Broken up 1928	
Lily of the Valley – FH 36, FH 27	Wrecked 1929	
Lewis – TO 50	Broken up 1921	
Lois	Broken up 1894	Clinker built
Lena	Falmouth	
Magpie 2		
Maggie Dixon		
Maud – FH 165	Left port 1970's	Clinker built. Had a centreplate
Murre	Left port 1979/80	
Magdalena – FH 43, FH 178	Falmouth	
Melita – TO 51, FH 186	Falmouth	
Mabel – TO 12	Sold to Germany about 1930	
Mayflower – FH 19	Falmouth	
Myrtle – FH 4	Left port	Probably burnt
Mildred – TO 20, FH 37	Falmouth	
Maiden Nellie – TO 30, FH 63	Rochester (Kent)	
Muriel – TO 42	Falmouth	
Moya	Left port ? 1980's	Morecambe Bay Prawner
Morning Star – FH 174	Lands End Leisure Centre	Became an exhibit after she was wrecked
Melody		Probably a carrier boat
Maria – TO 46	N.L.F. 1873	
Mary Jane		Ketch
Maria	Reg. cancelled 1878	
Mildred	Kiel, Germany	Now yacht with cabin renamed 'Turf'
Maid of the Mill – TO 87	Broken up 1871	
Mystery – TO 89		
Mary Ann – TO 94	Reg. cancelled 1874	
Majestic – TO 30	Crushed in accident	Sold as Yacht 1935
Margaret – TO 16		Sold as Yacht 1937
Minnie – TO 91	N.L.F. 1886	
Mollie	Falmouth	Now a yacht with cabin

Name	Present Whereabouts	Comments
Nellie – TO 70, FH 276	Holland	Still dredging until 1918
Odessa		Dandy
Ocean – FH 53	Broken up 1918	Dandy
Olga Joan – TO 1		Lug rig, clinker built. Sold as Yacht 1936
Obediah – FH 39		
Orlando – TO 16	Falmouth	
Onward – TO 2	Broken up 1932	Sloop, but used a staysail for dredging.
Olivia	Sank in the English Channel	Owners Jack and Tom Barron drowned when she was cut in half in the collision.
Orca	France	Now yacht with cabin
Pelican – FH 325	N.L.F. 1928	Lugger
Pet – FH 376	Falmouth	
Prosperity		
Pioneer – TO 85, TO 43		Spritsail. Possibly a carrier boat.
Pearl – TO 28	Condemned 1926	
P.C. – FH 170	N.L.F. 1927	Dandy. Altered to cutter 1912
Pearl		Sold to Cardiff as Pilot Boat
Pride of the Isle	Fell Apart at St. Just on the beach	Trennailed
Polly – TO 41		
Phantom		Built as Newquay Pleasure Boat
P.C.W. – TO 99		
Pandora – FH 238	Falmouth	Now yacht with cabin
Petrel – TO 35, FH 172	Broken up 1929	Counter stern. Smack Rig (large mainsail).
Queen of the Bay – TO 27, FH 151	N.L.F. 1932	Dandy. Sam Furse drowned from her when knocked overboard whilst out in Falmouth Bay 1905 mackereling.
Queen of the May – FH 181, FH 31	Fowey	Now yacht with cabin
Queen of the Fall – TO 92	Reg. cancelled 1874	
Rover – TO 88, FH 165	N.L.F. 1921	
Rob Roy		Oyster Carrier
Ripple – TO 48, TO 25, FH 303	N.L.F. 1934	Counter stern
Rambler – TO 86	Condemned 1880	

Name	Present Whereabouts	Comments
Rita	Falmouth	
Rebecca – TO 6	N.L.F. 1937	Dandy. Topmast
Result – TO 23	Bristol	Now a derelict
Royal Oak	Falmouth	
Rippling Wave – TO 20		
Serica – TO 31	Falmouth	
Softwing – TO 4	Falmouth	Owned by National Maritime Trust
Swallow – TO 7	Falmouth	Trennailed. Lengthened by William Searle at Pill Creek.
Shadow – TO 243, TO 10, FH 241	Falmouth	
St. Melorus	West Indies	Now yacht with cabin
Shag		Clinker built
Six Brothers (1) – TO 21, FH 133	Wrecked in gale 1969	
Six Brothers (2) – FH 24	Falmouth	
Skandinvia – FH 23	N.L.F. 1921	Yawl
Seabird – TO 10, FH 162		Clinker built. Sold as yacht 1937.
Sonny South		Built as Newquay Pleasure Boat.
Shamrock – TO 19	Falmouth	
Seacroft		
Swift – TO 1, TO 231, FH 234	Broken up 1919	
Stella – TO 24, TO 1	Falmouth	
Silent – FH 183		Had counter stern originally wrecked on passage to Ireland 1926
Sunbeam – TO 2		Now derelict
Serene – TO 29, FH 219	N.L.F. 1960	
Sprat	Condemned 1921	
Snowdrop – TO 32		Sold as yacht 1923
Susan Jane	Broken up at Point 1933	Trennailed. Nickname was 'Rigsabunky'
Shooting Star		
Seagull	Falmouth	
Sonny – TO 28		Ketch
Seacroft		
Twilight – FH 341	Scotland	Ketch. Had wheel steering
Torment		Clinker built
Three Brothers – FH 188	Broke up at Penzance 1973	

Name	Present Whereabouts	Comments
Twins – TO 96, TO 1	Lost 1887	
Thelma – TO 95, TO 83	Wrecked at Mylor 1962	
Teaser – TO 95, TO 83 FH 176	Lost 1923	
Thetis – TO 206	Reg. cancelled 1872	
Teazer		Counter stern
Two Brothers		Probably oyster carrier
Tern	Salcombe	Now yacht with cabin
Two Sisters – TO 11	Condemned 1928	
Three Sisters	Falmouth	
Unique		Caught fire and sank off Manacles. Similar lines to Mildred.
Veruna	Falmouth	
Vera – FH 72	Falmouth	
Vic – FH 12		
Vanguard		
Vivian	Helford	
Victory – TO 8	Falmouth	
Victoria		
Winkle Diver – FH 37	Falmouth	Yawl
Water Hen – FH 206	N.L.F. 1928	Had perfectly square forefoot.
White Heather – TO 13	Guernsey	
Wild Cat – FH 149	Sold to Topsham 1904	
William – TO 93		
William Green – FH 96	Sold as yacht 1933	
W.H.B.		Dandy
Winnie – TO 3	Falmouth	
William and John		
Willie TO 10		Ended her days under St. Just Church where her keel still lies.
Zephyr – TO 39	Falmouth	
Ziguener – FH 84	Broke up in Restronguet Pond 1951	
Zephyr – TO 16	Sold to Fowey 1927 as yacht.	Owned for one year by John and Henry Merrifield whilst they waited for Endeavour.
Zona – FH 67	Falmouth	

INDEX